TEACHING GUIDE
Geography of the Coordinate Plane

June Mark

E. Paul Goldenberg

Mary Fries

Jane M. Kang

Tracy Cordner

transition
to **algebra**

*make algebra
make sense*

Unit **6**

Geography of
the Coordinate
Plane

Research-based
National Science
Foundation-funded

EDC Learning
transforms
lives.

firsthand
HEINEMANN
DEDICATED TO TEACHERS™

*first*hand
An imprint of Heinemann
361 Hanover Street
Portsmouth, NH 03801-3912
www.heinemann.com

Offices and agents throughout the world

Education Development Center, Inc.
43 Foundry Avenue
Waltham, MA 02453-8313
www.edc.org

© 2014 by Education Development Center, Inc.

Co-Principal Investigators and Project Directors: E. Paul Goldenberg and June Mark

Development and Research Team: Tracy Cordner, Mary Fries, Mari Halladay, Jane M. Kang, and Josephine Louie

Contributors: Cindy Carter, Susan Creighton, Jeff Downin, Doreen Kilday, Deborah Spencer, and Yu Yan Xu

 This material is based on work supported by the National Science Foundation under Grant No. ESI-0917958. Opinions expressed are those of the authors and not necessarily those of the Foundation.

Transition to Algebra, Unit 6: Geography of the Coordinate Plane Teaching Guide
ISBN-13: 978-0-325-05320-2

Transition to Algebra Teacher Resources
ISBN-13: 978-0-325-05790-3

Transition to Algebra, Unit 6 Student Worktexts, 10-pack
ISBN-13: 978-0-325-05308-0

Transition to Algebra Student Worktexts, 10 Sets of All 12 Units
ISBN-13: 978-0-325-05791-0

Printed in the United States of America on acid-free paper

18   17   16   15   14   RRD   1   2   3   4   5

Unit 6

Geography of the Coordinate Plane

CONTENTS

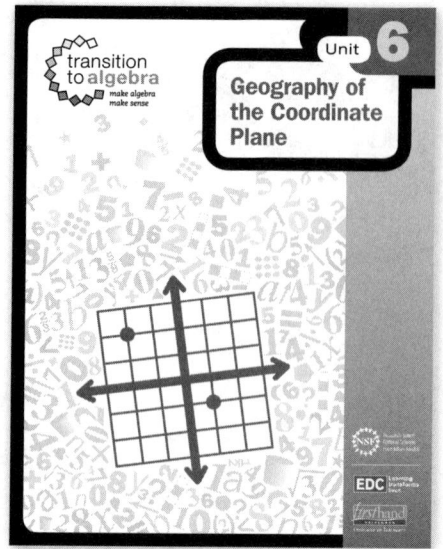

UNIT 6

Geography of the Coordinate Plane

Learning Goals

By the end of Unit 6, students should be able to:

- Understand coordinates on the plane as positions measured along two intersecting number lines.

- Observe and articulate relationships among data points on a graph (by comparing them or saying whether they generally follow a pattern).

- Describe and use transformations as functions that act on coordinates to produce new coordinates.

- Make and explain connections between stories and their graphs.

- Use equations as "point testers" (that is, use equations to determine whether a given point is a solution and therefore on the graph of the equation).

- Understand graphs as a collection of solution points.

- Generate a table of solution points and use it to create or identify a graph of an equation.

Unit 6 builds familiarity and facility with the conventions of the coordinate plane, graphs, and equation graphing. Students explore coordinated data to understand how a single point can represent (and connect) more than one piece of information. This understanding is furthered when students use transformations to affect a shape's x-coordinates, y-coordinates, or both. By the end of the unit, students graph equations using the idea that every point on a graph represents a particular x and y pair whose values are related by the equation. As students use equations to test arbitrary points to determine whether they are on the graph (and are solutions of the equation), they come to understand graphs as collections of solution points, which will support future work with graphs and equations in Unit 9: *Points, Slopes, and Lines*.

Seeing the Coordinate Plane as Perpendicular Number Lines

From Unit 2: *Geography of the Number Line* and Unit 3: *Micro-Geography of the Number Line,* students are familiar with locating a point along a one-dimensional number line. In this unit, students describe the location of a point on a plane using two intersecting number lines, the coordinate axes. Two numbers—the horizontal and vertical positions—locate each point; conversely, a single point represents two coordinated pieces of information.

Students sometimes confuse the order within the ordered pair or the names of the axes. Establishing that the first number goes on the horizontal number line (like before) and the second number, which represents the second dimension, goes on a second, new, vertical number line helps students remember this distinction and further connects the logic of the coordinate plane to students' established knowledge of the number line. Negative numbers and rational numbers continue to have the same meaning along an axis as they do on a number line, and students can extend the same logic they have built about relative placement and order of numbers to this new context.

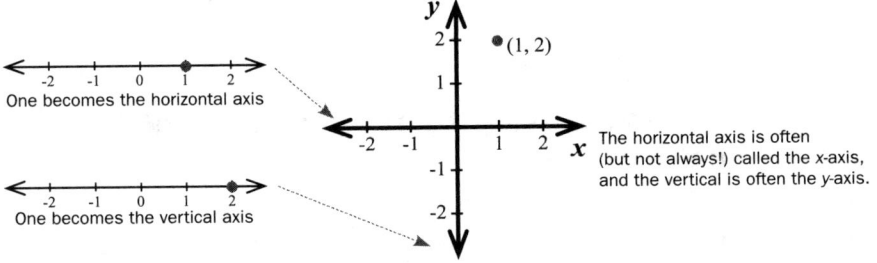

One becomes the horizontal axis

One becomes the vertical axis

The horizontal axis is often (but not always!) called the x-axis, and the vertical is often the y-axis.

Transformations

Students perform transformations on sets of points to explore translation, reflection, and dilation, to increase their facility with the coordinate plane, and to support future work with functions. Transformations behave like functions in that they take an input—in this case, a point on a plane (represented by a pair of numbers)—and output another point, determined by some rule. Transformations also support intuitive thinking about performing operations on graphs. For instance, if we know what $y = \frac{1}{3}x$ looks like, then we know something about what $y = \frac{1}{3}x + 5$ looks like.

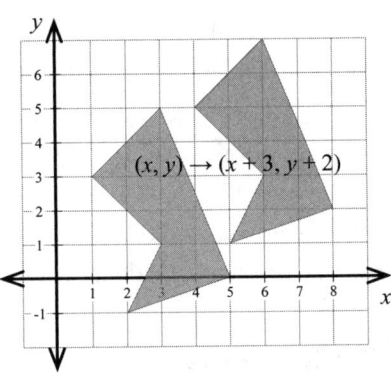

$(x, y) \rightarrow (x + 3, y + 2)$

Point Testing

Every point on an equation's graph is a solution to the corresponding equation. To support students' understanding of this essential idea, equations are treated as "point-testers" that algebraically determine whether a given point is on the graph. Students examine an equation like $y = 2x + 3$ from different perspectives. They treat the equation as a clue to a puzzle and see how, together with another clue like $x = 5$, it determines a single solution point $(5, 13)$, a point that is simultaneously a solution to *both* equations. Students are also presented with an assortment of points—for example, $(0, 3)$, $(-4, -11)$, $(3, 10)$, and $(-10, -17)$—and are asked to identify which of these are solution points for the equation $y = 2x + 3$. Students also generate their own solution points and use them to construct a graph of an equation. All of these perspectives are meant to help students establish both intuitive and precise connections between an equation, a table of solution points, and the graph of the equation. This approach allows students to extend beyond linear equations in slope-intercept form; students will use the same understanding to examine equations like $x + y = 4$ and $y = x^2 - 3$ and, later in Unit 9: *Points, Slopes, and*

Lines, $y^2 + x^2 = 25$. Their understanding of solution points and equations as point-testers will also support later work in graphing and solving equations by preparing students to see a graph as a collection of points in a pattern described by an equation.

Mental Mathematics: Using Approximations to Make Exact Calculations

As students explore the coordinate plane, they develop ideas about how distance works in this space. In Unit 6, students primarily experience distance on the coordinate plane along either the x-direction or the y-direction as they transform shapes or consider horizontal and vertical distances between points. (Distance between any two points on the plane [found using the Pythagorean equation $a^2 + b^2 = c^2$] will not be explicitly addressed until Unit 9: *Points, Slopes, and Lines.*) Finding distance along one dimension is supported by the Mental Mathematics activities of finding the distance between two numbers, such as a number and 30 (or 29, or 100). Students further use their mental strategies for finding distance as they add and subtract numbers like 9, 8, 19, 18, or 99. As students build a strategy for adding 19 (perhaps by first adding 20 and then subtracting 1), they are also making sense of the relationship between expressions like $x + 19$ and $x + 20 - 1$. This thinking supports work in Unit 6, where students will continue to make sense of equations as processes that encode patterns of calculation.

Explorations

In the Boxes of Chocolates Exploration, students examine a pattern of chocolates and explore possible groupings that make it easier to find the total number of chocolates without counting each one. This supports their use of structure and their ability to connect a geometric object to numerical patterns and an algebraic expression.

In the Street Paths Exploration, students draw paths along grid lines on the coordinate plane and measure these paths between specified points. This builds students' skills with identifying and describing patterns and supports two-dimensional reasoning.

Related Game

The game Battleshape gives students an enjoyable way to practice naming coordinate pairs in the context of the geometry of shapes on the coordinate plane.

Lesson 1:
Plotting Data

PURPOSE

By plotting individual data points before plotting the graphs of functions, students see the meaning that can be captured in a point. Then, when students later graph equations (like $x + y = 7$) in which the points pattern themselves into lines and other curves, the individual points retain their meaning as solutions to the equations. In this lesson, students practice plotting ordered pairs and see the relevance of plotting information along two coordinated number lines at once. Where Am I? puzzles provide practice plotting points and deducing location on the coordinate plane. These puzzles will continue to appear throughout the unit as students work up to graphing equations.

 **Mental Mathematics** *Begin each day with five minutes of Mental Mathematics (pages T51–T64). This unit's activities focus on finding the distance between two numbers and using approximations to make exact calculations.*

Launch: Arm Span versus Height

Have students work together to measure arm span and height (in centimeters) and record that information on a sticky note as an ordered pair, listing arm span first as shown below. Make sure they write their name fairly large, as you will be asking questions about individual sticky notes and you want students to be able to see the names from their seats.

Next, display the arm span number line (page T38) and have students place their sticky note where it belongs. After all the sticky notes have been placed, check for understanding. Select a sticky and ask, **What is this student's arm span?** or ask **Who has the smallest arm span? The largest?**

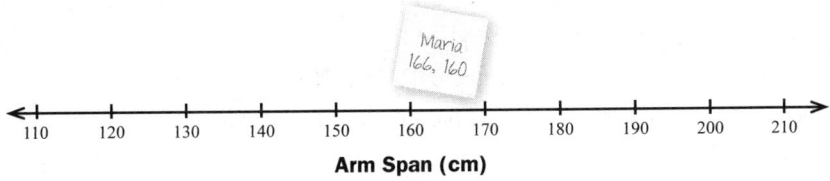

Arm Span (cm)

Display the arm span vs. height graph on page T39 in the same place, aligning the horizontal axis with the sticky notes from the previous activity.

Invite students to move their sticky to properly represent their height information on the coordinate plane by moving their sticky *straight up* (preserve its horizontal position). When everyone has found their place on the

Lesson at a Glance

Preparation: Students will need some way to measure their arm span and height *in centimeters:* measuring tape, measuring sticks, or just long strips of paper or string that can be measured with a ruler. If possible, create measuring "stations" with meter sticks (or measuring tape) taped to the wall (some horizontally and some vertically) to help students collect accurate measurements.

- Each student will need a sticky note to record and label their data.
- Prepare to display the blank number line on page T38 and the blank graph on page T39.

Mental Mathematics (5 min)

Launch: Arm Span vs. Height (25 min)

- Students measure and record their arm spans and heights on two intersecting number lines (the coordinate plane).

Student Problem Solving and Discussion (15 min)

- Allow students to work through the Important Stuff and explore additional problems.
- Ask students to share their responses to the *Discuss & Write* problems about hypothetical arm span and height scenarios.
- Discuss the meaning of data represented as points on graphs.

Unit 6 Related Game: Battleshape (See page T37 and Student Worktext page 43.)

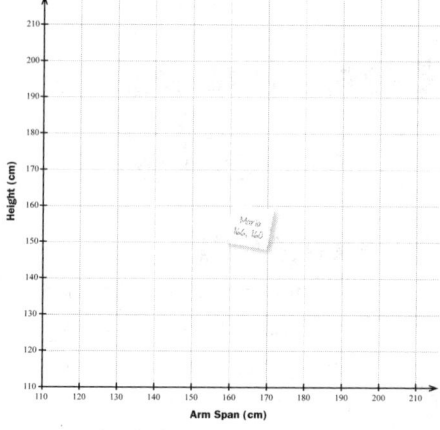

Algebraic Habits of Mind

Communicating with Precision

In this lesson, students experience graphing as a way of communicating information. Interpreting points as representing and associating two pieces of information provides a *reason* for being consistent about the order in which we list those two numbers and how we use them to plot the points on a graph. To graph arm span versus height, students assign meaning to points (each point represents one person's data) and to their coordinates (*everyone* must agree that the first coordinate corresponds to arm span and that arm span will be represented on the horizontal axis). As students develop the habit of mind of looking for the information communicated by a graph, they start to understand why graphing is such a powerful tool for representing and analyzing mathematical patterns.

In this lesson, we refer to the axes as "arm span" and "height" (or horizontal and vertical), not as *x* and *y*, because there are no *x*'s or *y*'s in this context. Students still label the points with ordered-pair notation. Encourage them to associate the first measurement (arm span) with the horizontal axis, which is just like the number line with which they are already familiar. The second, new measurement (height) then goes on the new, vertical axis.

coordinate plane, ask students to interpret the plotted information. Use prompts like "How tall is Joe? What is his arm span? Who has the widest arm span? Is he/she also the tallest? Who has a 150-cm arm span? What is the arm span of the person who is 156 cm tall?"

Ask students: "From the appearance of the graph, does arm span seem in any way related to height? (How do we see that without reading off the numbers?) Is that what we might expect of other classrooms as well? How would the graph look if height and arm span were completely *un*related?" The data are likely to indicate that height does seem to increase along with arm span. Listen for reasoning that identifies both height and arm span as being related to the overall size of a person, so it makes sense that they would be related. If the two were unrelated, the points on the graph would be scattered without any apparent pattern.

Consider using these additional prompts to discuss graphs:

» **If everyone in our class had the exact same arm span, what would our graph look like?** With arm span on the horizontal axis, all the data points would fall along a vertical line.

» **If everyone's arm span exactly equaled their height, what would our graph look like?** The graph would show points that lie along the line *height = arm span*. Have students figure out what the graph of this equation would look like (a 45° diagonal going through the origin to the upper right).

» **Does it make sense to connect points in our arm span vs. height graph?** It might make sense to draw a line that shows the general pattern of the relationship between arm span and height (the line of best fit), but drawing this line isn't simply a matter of connecting points. This line might not even go through a single data point from the class, so it is actually inaccurate for describing any specific person's height and arm span relationship. But this line can be calculated to show an *average* relationship between height and arm span given the class as a whole. The points along that line have meaning (as the *predicted* height for a given arm span), although the points may not represent any student in the class.

Here are some suggestions to extend this activity and deepen discussion:

• Slide a measuring stick along the graph perpendicular to the arm span axis to observe the order of arm spans. Repeat this action perpendicular to the height axis to observe the order of the heights.

- You may wish to experiment with moving a sticky note around to represent improper ways to arrange the data and ask students to *explain why each will not work*. For instance:
  - Move a sticky up or down, keeping the arm span correct: "Does this point still represent Joe correctly?"
  - Move a sticky left or right, keeping the height correct.
  - Replot one student with height and arm span swapped.
- Make a list of "Students Ordered by Arm Span" from shortest to longest arm span. Comment that the numbers give a "one-dimensional picture" of the class based on just one dimension of information about each student, which can be ordered forward or backward, but that's all. This one piece of information can be arranged along a one-dimensional number line.
- Then make a list of "Students Ordered by Height" (again starting with the shortest) and compare that to the first list. If the lists are identical, you may consider making up a couple of fictitious students to distinguish the lists. Observe that there are now *two* ways of ordering the information: by arm span or by height.
- Observe that measuring *two* dimensions of information about people (in this case, arm span and height) means that we need *two* axes (one for arm span and one for height) to plot that information. There are now two ways to order the information, and we could line all the stickies along either of two lines. Or, as we just did, we could *coordinate* both pieces of information and spread the points out over the plane.
- Ask if there are *other* ways to order the information. If students suggest alphabetical, observe that this could be thought of as a *third* dimension, and the stickies could be moved along a third axis, perpendicular to the board, marked with the letters of the alphabet. That is exactly the right idea about dimension: it is an independent piece of information.

Student Problem Solving and Discussion

PROBLEMS 1–3 of the Important Stuff may be done in conjunction with the Launch activity.

PROBLEMS 4–5 in the *Discuss & Write* box ask students to consider a hypothetical data point and a hypothetical situation related to the height vs. arm span graph. Listen for student responses indicating that they understand that the horizontal and vertical axes show different, independent measurements.

Have students solve the Where Am I? puzzles in PROBLEMS 6–12, working together if they like. The puzzles are new and may require some explanation, but as much as possible, let students figure out on their own how to proceed. The first mention of *x* and *y* occurs here. Problems 6–8 are plotting questions. Problems 9–12 require students to think of the *y*-coordinate of a point in terms of its relationship to its *x*-coordinate.

Lesson at a Glance

Mental Mathematics (5 min)

Launch: Coordinating Data (10 min)

· Given only two lists of five people ordered by arm span and by height, students create a possible graph of the data. Students consider each dimension independently and then the two together as coordinated data points.

Student Problem Solving and Discussion (30 min)

· Allow students to work through the Important Stuff and explore additional problems.

· Have students share responses from the *Discuss & Write* problems about the information presented in the graphs.

· Discuss what it means to coordinate data and what can be said about relationships by looking at graphs.

Unit 6 Related Game: Battleshape (See page T37 and Student Worktext page 43.)

Lesson 2:
Coordinating Data

PURPOSE

Unlike a list of data, graphs present information in a way that can make patterns visible. This lesson helps students build facility with interpreting data on a coordinate plane and supports understanding of how sets of ordered pairs represent coordinated data. Students also encounter a situation where two measurements (time spent and grade received) are known but the relationship between them (if any) is not known and a graph can help discern the relationship.

Lessons 1 and 2 help build understanding that the geography of the coordinate plane is all about the relationship between two distinct dimensions. In Unit 9: *Points, Slopes, and Lines,* students will find the *distance* between two points on the plane and the *slope* of a line between two points on the plane. Both concepts of distance and slope require students to measure the *x*-distance and *y*-distance separately. Both of these ideas require the ability to distinguish a change in the horizontal direction from a change in the vertical direction.

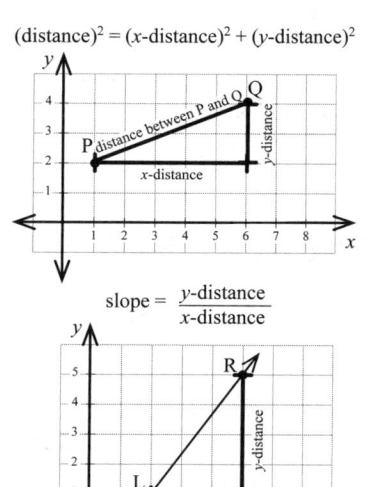

 Mental Mathematics *Begin each day with five minutes of Mental Mathematics (pages T51–T64). Encourage students to picture the number line when finding distances. Connect these distance-finding activities to the unit by describing the coordinate plane as two coordinated number lines.*

Launch: Coordinating Data

Make two lists of five students: one that orders the students by arm span and another that orders the students by height. Do not make the lists identical, but vary the order. For example, for students A, B, C, D, and E, you may order their arm spans B < E < A < D < C and their heights A < D < E < B < C.

Explain to students that it is *arbitrary* which axis to choose for arm span and for height. Sometimes the decision is not arbitrary, but in this case there is no "independent" or "dependent" variable. We decide on an order and all follow that convention, not just to be alike, but so that we can discuss and compare our results without confusion. In the graphs shown, arm span is on the horizontal axis.

Have students consider individually, then with a small group, what a graph for the five students might look like. In a whole group discussion, ask students to share strategies for how they *started* their graph. Listen for the strategies your students find.

Three possible strategies

Strategy 1: Begin by choosing a point (such as A), plotting it (so that its arm span is in the middle and its height is low), and then choosing another point (B) and plotting it relative to A (in this case, shorter arm span, greater height). Repeat, placing other points relative to these.

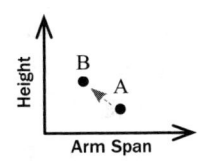

Strategy 2: Consider only arm spans and plot them along the horizontal axis as on a number line (in order, B, E, A, D, C), and then move the points up in order from shortest to tallest (i.e., start by plotting A, then plot D so it's higher than A, etc.).

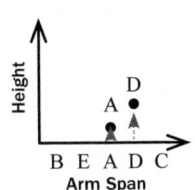

Strategy 3: Consider arm span and place B, E, A, D, C along the horizontal axis. Then place the letters A, D, E, B, C along the vertical axis. Coordinate this information to plot a single point for each letter using the corresponding horizontal and vertical positions marked.

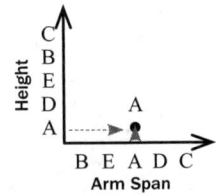

No matter what strategy students use, they must juggle two independent pieces of information to graph the points. Help students see that they are *coordinating* two pieces of information, and points that coordinate two pieces of information are graphed on a coordinate plane.

Student Problem Solving and Discussion

After most students have completed the Important Stuff section, ask them to share responses from the five *Discuss & Write* questions. For **PROBLEM 4**, listen for responses indicating that students appreciate how representing data on the coordinate plane makes it possible to see two pieces of information simultaneously. For **PROBLEMS 6 & 7**, listen for students who see that the relationship between time spent online and grade received is not straightforward (i.e., it's not the case that students who studied most online always got the best grades). Listen for details that justify their description of the relationship between hours and grade. You may ask students to put forth hypotheses for the relationship (perhaps it's because *how* you study matters—it is not just a matter of time spent, but how effective the studying is).

Problems 6 and 7 highlight two ways of "seeing" points on a graph. The first way is to use graphs to sort out and examine individual data points. Problem 6 uses this idea when it asks students to pick out students who have spent the most time online and examine their grades. The graph helps to make this information accessible without going through the disorganized table of values. The second way of "seeing" points on

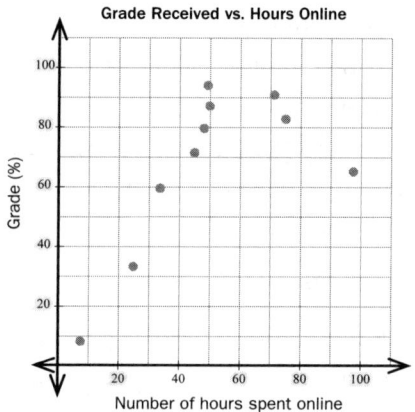

Grade Received vs. Hours Online

a graph is to examine how the graph makes an overall relationship visible. In discussing problem 7, listen for and address students who seem to be too caught up in individual data points to be able to see the overall relationship.

Discuss & Write What You Think

Graphs communicate a lot of information without using a lot of words.

(6) Did the students who spent the most time online also get the highest grades in the class? Explain. (Responses will vary.)
No. The person who spent the most time spent around 100 hours online. But that student scored almost 30 points lower than the student who got the highest grade, who only spent around 50 hours.

(7) In general, does the graph show a relationship between amount of time studying online and grade in the course? Explain. (Responses will vary.)
Yes. Those who spent very little time studying online got poor grades. Those who studied more got higher grades. But the effect drops off after 60-70 hours and those who studied beyond that did less well.

For further discussion:

» **For the MP3 graph in problem 1, does it make sense to connect those points? Why or why not?** Here it does not make sense to connect the points. It's hard to even tell in what order the points would be connected. Also, the lines between points would have no meaning.

» **Which MP3 player would *you* be most likely to purchase?** This is a matter of opinion, but listen for how students use the graph as they choose their preferences. If we choose just on the basis of most storage per dollar spent, the Creativity appears to be the best deal, but the Fony and Dansa also have more storage space compared to brands with similar prices. Students' criteria for "best" may vary. Keep referring students back to the graph. Ask them, **Where is this information on the graph?**

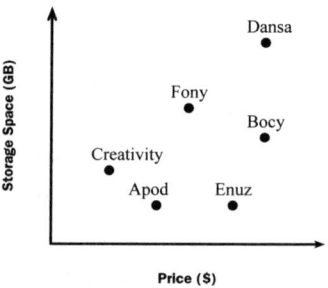

» **For problems 5–7, would you be able to answer those questions with the table alone? How does the graph help?** It is possible to answer these questions using the table alone, but the graph makes the answers visible without as much work. With the graph, it is possible to see the order of the points separately in both dimensions (time and grade) and together as coordinated data.

» **For the graph in problem 5, does it make sense to connect those points? Why or why not?** It doesn't make sense to connect the points with short line segments that go from point to point. There would be no meaning to most parts of such lines. Like the graph of arm span vs. height from Lesson 1, it might make sense to draw a graph that represents the general or average pattern (the line of best fit), but even this is unclear. Is the relationship truly a curve, or is it actually a straight line relationship and Student 2 just happened to do very poorly in the class? This can't be answered from the graph as shown but may be clarified by looking at more data points. It is not important that all students grasp such subtleties of this question right now, but these points may come up in discussion. The important idea for students to understand is the value of considering the meaning of information plotted on a graph. As these points represent individual data, a line of best fit could be used to predict how other students may score but will likely not match the data accurately. Lines or line segments simply connecting the points often do not have a useful meaning.

Lesson at a Glance

Mental Mathematics (5 min)

Launch: Think of a Point (10 min)

· Discuss how a given transformation affects every point in a similar way.

· Have students go through examples, make a general statement, and express the transformation algebraically.

Student Problem Solving and Discussion (30 min)

· Give students time to work through the Important Stuff and explore additional problems.

· Have students share responses to *Discuss & Write* problems, which ask students to reason about a transformation and address a common misunderstanding.

· Check for student understanding of the algebraic notation used to represent these transformations.

Unit 6 Related Game: Battleshape (See page T37 and Student Worktext page 43.)

The notation used in transformations, such as $(x, y) \rightarrow (x, y + 4)$, is a shorthand way of describing what happens to *any* point (x, y). Help students continue to make connections between verbal descriptions and algebraic expressions to make sense of this notation and to strengthen their understanding of variables.

Lesson 3:
Geometric Transformations

PURPOSE

This lesson shows how algebraic language can be used to describe some geometric ideas, giving students practice with evaluating expressions in a new context. In transformation problems, students use algebra to change a collection of points. In order to do this successfully, students must evaluate expressions, work with functions that describe transformations, and keep track of each value in an ordered pair separately. In this lesson, students focus entirely on translation transformations that do not change the size or orientation of the original shape.

Transformations are functions: something goes in (in this case, a point), a change is made "inside the box," and something comes out (another point). The notation $(x, y) \rightarrow (2x, y - 5)$ is used instead of "standard" function notation, such as $f(x)$ or $f(x, y)$, to avoid introducing new letters and a new meaning for parentheses. You don't need to discuss function notation in this unit, but it is important to emphasize that transformations transform every point in a graph.

$(5, 3)$

Transformation
$(x, y) \rightarrow (2x, y - 5)$

$(10, -2)$

Proper understanding of transformations in this unit supports proper understanding of functions later. For example, working with transformations, such as $(x, y) \rightarrow (x, y + 9)$, helps prepare students for understanding and identifying similarities and differences between graphs of different equations, such as $y = x$ and $y = x + 9$.

 Mental Mathematics *Begin each day with five minutes of Mental Mathematics (pages T51–T64). In the lesson, students will treat geometric transformations as functions whose inputs and outputs are points. In Mental Mathematics activities, students also enact functions, receiving inputs and producing outputs according to a rule.*

Launch: Think of a Point

This activity is a variation of the Think of a Number tricks that students first explored in Unit 1: *Language of Algebra*. To begin, draw a blank table on the board:

Think of a point.							

Instruct students to "Think of a point." Students should choose a point they can picture on a coordinate plane in their minds. Tell students, "Now move the point up 4 units." Write the instruction on the table. Ask students to share their examples and record these examples on the table. Your table might look like this:

Think of a point.	(2, 4)	(-4, 1)	(-13, 0)	(3, -3)	(-1, -8)	$(\frac{1}{2}, 3\frac{1}{2})$	(x, y)
Move up 4.	(2, 8)	(-4, 5)	(-13, 4)	(3, 1)	(-1, -4)	$(\frac{1}{2}, 7\frac{1}{2})$	

Ask students, **You know how to find the output if the input is (2, 4) or (3, -3). What is the output if the input is (x, y)?** If students answer right away, record their answer. If students are unsure, focus their attention on the similarity in the treatment of each input in the table. Ask students to describe in words how to find the new point. When students say something like "The first number stays the same, and we add 4 to the second number," capture those words with variables: "The first number is x, so x stays the same, and the second number is y, and we add 4 to y." Fill in the last space with $(x, y + 4)$.

Student Problem Solving and Discussion

Give students time to work, independently or in groups, on the Important Stuff. Bring the class together to discuss PROBLEMS 4 & 9 in the *Discuss & Write* boxes. Problem 4 asks students to focus on the independence of the x- and y-coordinates.

(4) Explain why it makes sense that the x-coordinates of the vertex points should stay exactly the same when the heart moves down.

In responding to problem 9, students communicate their understanding that this kind of transformation (a slide transformation using only addition or subtraction) does not change the size of the shape.

(9) A classmate says, "When I look at $(x, y) \rightarrow (x + 15, y + 30)$, I think that means the shape becomes bigger. See, we're adding numbers to both x and y. Doesn't that mean the shape gets larger?" Discuss and write how you would explain to this student what's really going on with the shape.

When a constant value is added to every x-coordinate, such as in the transformation $(x, y) \rightarrow (x + 8, y)$, the x-coordinate of *every* point is affected, so the effect is a horizontal change in location. In Lesson 4, students will also encounter transformations that have a multiplicative effect. In the case of a transformation like $(x, y) \rightarrow (8x, y)$, the size of a shape might be affected because the distance between x and $8x$ is $7x$, which depends on the magnitude of x, whereas the distance between x and $x + 8$ is always 8 and doesn't depend on the magnitude of x. For now, it is enough for students to understand that transformations involving adding or subtracting a constant value affect only the location of a shape.

Gauge your students' understanding of transformations and their facility with algebraic representation with questions like these:

» **Imagine that you've made a shape and you want to repeat it across a page, every 10 units. Describe the transformation in terms of what happens to the points of your shape. How could algebraic notation make your description easier?** The points in the first copy would have the same y-coordinates as the original image and 10 added to each x-coordinate of the original. The second copy would again have unchanged y-coordinates and 20 added to each original x-coordinate, and so on. Using algebra, the first copy would be made with the transformation $(x, y) \rightarrow (x + 10, y)$, then the second with $(x, y) \rightarrow (x + 20, y)$, the third with $(x, y) \rightarrow (x + 30, y)$, and so on. Using algebraic notation even further, the nth copy would have coordinates according to the rule $(x, y) \rightarrow (x + 10n, y)$.

Formal vocabulary is not the focus here. Students may use creative language to describe the transformation they are working with: "move," "slide," "push," or "go," or perhaps even "translate." Accept their answers as long as the mathematics is accurate and the language is appropriate for the transformation.

» **When you see a transformation written algebraically, how do you know if it will move a figure down? Or left? Or up and to the left?** Listen for students who have made sense of the idea that a transformation right or left affects the x-coordinate and a transformation up or down affects the y-coordinate. Furthermore, adding a positive number moves a figure to the right or up, while subtracting moves a figure to the left or down.

» **On a graph, if x-coordinates were arm spans and y-coordinates were heights, what would $(x, y) \rightarrow (x + 3, y)$ represent?** This could represent a situation where all the arm spans grow exactly 3 centimeters while the heights stay the same. More realistically, this transformation might be necessary if, after you have done all the measurements, it is found that the measuring stick used for arm span was 3 cm off. Pose similar questions connecting transformations to data. For example, in the MP3 graph (Lesson 2), what would $(x + 30, y + 50)$ represent? (This may be what happens as technology progresses—the price goes up by $30, but the data storage space goes up by 50 GB for the same brands of MP3 players.) Of course, this scenario is not realistic, but it can help students to make sense of the meaning of such a transformation in context.

» **In the tables, we kept track of the shape by keeping track of corner points. What would happen if we chose to examine points in the shape that were not at the corners?** The corner points are convenient, but students should see that *all* points in the shape (on the boundary or inside) are transformed in the same way by a transformation using addition or subtraction (a translation). In fact, knowing that translations preserve size and orientation, students only have to transform one point to be able to draw the rest of the shape—keeping track of *all* the corner points is not necessary. For example, in problem 1, just from knowing that the point (1, 1) is transformed to (6, 1) (and knowing that the transformation is a translation), it is possible to draw the entire transformed triangle.

Lesson 4:
Transformations with Algebra

PURPOSE

The transformations in this lesson involve multiplying one or both coordinates by a factor. The magnitude and sign of the factor indicate the size and orientation of the new image. Working with reflections and dilations gives students a visual demonstration of multiplication by positive and negative numbers. Multiplying a coordinate (such as the x-coordinate) by an integer with magnitude greater than 1 stretches the figure away from the other coordinate's axis (in this case, the y-axis, where $x = 0$). Multiplying by a number between -1 and 1 compresses the figure towards that axis. These effects are consistent with those of multiplying by a number with magnitude greater or less than 1 on a number line. In any case, multiplying a coordinate by a negative number reflects the figure across the other coordinate's axis, just as points on the number line move to the other side of zero when their sign is changed.

The exact vocabulary involved with transformations is not the purpose of this lesson. The purpose is to provide students with experience reading the coordinates of points, evaluating algebraic expressions using the coordinates' values, and plotting and interpreting the results. This gives students useful practice in basic graphing skills while continuing to build experience with functions.

 **Mental Mathematics** Begin each day with five minutes of Mental Mathematics (pages T51–T64). Continue to encourage students to view these activities as a type of mathematical transformation.

Launch: Reflecting Shapes

Go over Transformation 1 of **PROBLEM 1**. Have students fill out the table and draw the figure (it will be reflected across the x-axis). Let them work through Transformation 2 on their own, and discuss what they notice. This brief activity introduces transformations involving multiplication instead of addition. Encourage students to make sense of the transformations by connecting the algebra and effect on the pictures. Ask questions like **What happens to the shape? What part of the algebraic rule explains what happens to the shape?** For example, it makes sense to indicate reflection across the x-axis with $(x, y) \rightarrow (x, -y)$ because the x-values must stay the same and the y-values take on the opposite value. Use the language of "opposite" rather than "negative," because as students reflect the shape across the y-axis, this may involve finding the opposites of negative values.

Lesson at a Glance

Preparation: Photocopy the Snapshot Check-in on page T45.

Mental Mathematics (5 min)

Launch: Reflecting Shapes (5 min)

· Students explore and discuss problem 1 in the Student Worktext, which reflects a shape over the x-axis and over the y-axis.

Student Problem Solving and Discussion (25 min)

· Have students work through the rest of the Important Stuff and explore additional problems.

· Have students share responses to the questions in the *Discuss & Write* boxes, and then further discuss the difference between the effects of addition and multiplication in transformations.

Reflection and Assessment: Snapshot Check-in (10 min)

Unit 6 Related Game: Battleshape (See page T37 and Student Worktext page 43.)

Watch for difficulties that indicate misunderstandings (like getting the signs or variables confused, plotting points incorrectly, or losing track of what exactly is being read, processed, and recorded), and guide each student to focus on improving specific skills.

Seeking and Using Structure

Students seek and use structure when they compare points, use visual clues to describe a transformation, and express that transformation with algebra to show how the process can be applied to *any* point.

A lot of thinking goes into a task like this. Students must compare shapes before and after the transformation and recognize which parts correspond to each other (which can be challenging when the shape has been distorted or reflected). In order to produce a generalized transformation rule, students must understand that it is possible to look at the visual effects of a transformation for clues about the rule. For example, a shape that has been reflected over an axis is a clue that the transformation rule involves a change in sign.

Students develop the habit of mind of seeking and using structure when they use observations about points (e.g. they notice that every *y*-coordinate has doubled) to write a transformation rule like $(x, y) \rightarrow (x, 2y)$ and see that the visual effects of the transformation supports their thinking.

Words students are likely to use to describe these transformations include "flip," "reflect," "stretch," "shrink," "fatten," "squish," "enlarge," and "reduce." Encourage any language that reflects correct mathematical thinking.

Student Problem Solving and Discussion

Let students work on the Important Stuff problems on their own or in groups.

PROBLEMS 6–8 ask students to generate a transformation rule by observing shapes before and after a transformation. These problems practice the habit of mind of seeking and using structure. Students may also engage in the habit of mind of describing repeated reasoning as they record specific examples of points that have been transformed and then write the general rule and make sense of it.

Encourage collaboration on **PROBLEM 9**, which asks students to invent a transformation to make a shape "taller but not wider." Listen for students who realize they can consider what happens to the *x*-coordinates separately from the *y*-coordinates. You may also hear students wonder if they can create a transformation that will cause the shape to not fit on the limited size of the grid. This is good! It shows that students are thinking about the grid as only one small part of the infinite coordinate plane. You could offer a larger piece of graph paper if they want to try their ideas.

Ask students to share their responses to **PROBLEM 12** in the *Discuss & Write* box. Listen for responses indicating recognition that multiplication must be involved, since the size of the shape changes, and that it is the *x*-coordinate that is scaled. Make sure students are precise in their answers and specify why the *x*-coordinates must be multiplied by a number greater than 1. Ask students to describe what might happen if the *x*-coordinates are multiplied by a number less than 1, such as $\frac{1}{2}$.

Use these questions to compare the effects of addition and multiplication in transformations. Provide examples of shapes (including some that cross the axes) that students can examine together as they explain their thinking.

» **What happens to a shape under the transformation rule $(x, y) \rightarrow (2x, y)$?** Each *x*-coordinate is doubled, which means that each point is twice as far from the *y*-axis as it was originally. The shape becomes exactly twice as wide as it had been (with width measured horizontally).

» **Will a shape transformed by $(x, y) \rightarrow (2x, y)$ always end up to the right of the original?** It is *not* true that the shape *must* move to the right; this is true only for points with a positive *x*-coordinate. If the *x*-coordinate of a point is 0, the point does not move at all, and if the *x*-coordinate of a point is negative, the transformed point ends up to the left of the original point.

» **What is the difference between the transformation $(x, y) \rightarrow (x, y + 5)$ and $(x, y) \rightarrow (x, 5y)$?** In the first transformation, the shape moves up 5. The transformed shape is exactly the same shape and size as the original. In the second transformation, every point is five times as far from the *x*-axis as in the original. The shape has been stretched so that its height (measured vertically) is five times what it was. The width remains the same, though, so the relationship between the shape's width and height has changed.

Student Reflections & Snapshot Check-in

Ask students to reflect on their learning:

What are some things you've learned so far in this unit?

What questions do you still have?

Assess student understanding of the ideas presented so far in the unit with the Snapshot Check-in on page T45. Use student performance on this assessment to guide students to select targeted Additional Practice problems from this or prior lessons as necessary.

So far in Unit 6 students have:

- Explored points on the coordinate plane—both as pairs of coordinated data and as points in a shape to be transformed.
- Graphed points (by locating the first coordinate along a horizontal number line and then adjusting that point according to the second coordinate using a second, vertical number line).
- Coordinated data, seeing one point as a way to relate two measurements, such as the arm span and height of a person.
- Compared points by comparing *x*-values or *y*-values independently and together (e.g. comparing price and storage space or price per unit of space).
- Moved shapes up or down (by adding to or subtracting from *y*-values) and/or right or left (by adding to or subtracting from *x*-values).
- Made shapes larger or smaller and/or flipped them across axes by multiplying their *x*- and/or *y*-values.

Students have been developing the following Algebraic Habits of Mind:

- **Communicating with Precision**—Students have used the labels on the axes, the relative positions of points, and any regularity or pattern in the points to figure out what information is communicated by a graph.
- **Seeking and Using Structure**—Students have seen how transformations applied to a shape affect every point of the shape in a way that can be described generally with an algebraic expression.

· If necessary, suggest vocabulary like "speeding up," "slowing down," "fastest/slowest speed," and "original speed."

· The labels on your class's picture may differ from this example based on students' level of experience with skateboarding. The precise labels (such as whether the skateboarder begins the descent at a slow speed or at the original speed) are not as important as the general idea of speeding up and then slowing down and using this information to construct an accurately representative graph of speed over time.

Lesson 5: Intuitive Graphing

PURPOSE

A sketched graph can be used to express the general *relationship* between two quantities without specifying either quantity precisely. In this lesson's Launch, the two quantities are the *speed* of a skateboarder and *time*. A precise graph could be generated from an experiment, but even without data, we can sketch a graph that expresses the relationship well.

A graph can show a story. In this lesson, students will analyze graphs and stories to decide what's important to represent. The purpose of this lesson is to engage students in capturing the complexity of a story using a graph.

 Mental Mathematics *Begin each day with five minutes of Mental Mathematics (pages T51–T64). Students continue to perform more difficult addition problems (like adding 19 and 18) by first adding a simpler number (20) and then adjusting.*

Launch: Skateboarder Speed

Display the diagram (on the top of page T41) showing a skateboarder on a curved structure (half-pipe). Describe the scenario together—a skateboarder approaches, takes the board down and then up the structure, and finally skateboards away. Ask students to describe the *speed* of the skateboarder at the various times in this trip.

Invite students to label the skateboarder's speed directly on the picture, perhaps like this:

The skateboarder was going at some speed and then slowed at the start of the half-pipe. The speed increased on the way down and was fastest at the bottom. On the way up the other side of the half-pipe, the skateboarder slowed down again to a complete stop at the top of the half-pipe. The skateboarder then returned to the original speed.

After students agree on a story, have the class think about what a graph of the skateboarder's speed over time might look like. There are no numbers, but there is enough information to reason about the shape of the graph.

In this case, the mental picture of the skateboarder's *position in space*—the picture of the half-pipe—can draw students' attention from the task of creating a picture of the skateboarder's *speed over time*. The graph of speed is not the same shape as the path on the half-pipe. Help students focus on using the words and ideas they described earlier ("decreasing," "slowest," etc.) to guide how they record speed on the new graph.

If students need help starting, you might suggest that they think about just the time axis. Students should see that going down the half-pipe takes less time than coming up. So students might label the following along the time axis:

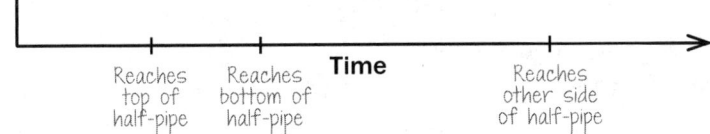

Then students can use these marks to plot points showing the relative speed of the skateboarder at those times. Engage students in discussing what the graph should look like as the skateboarder speeds up and slows down. A final graph may look something like this:

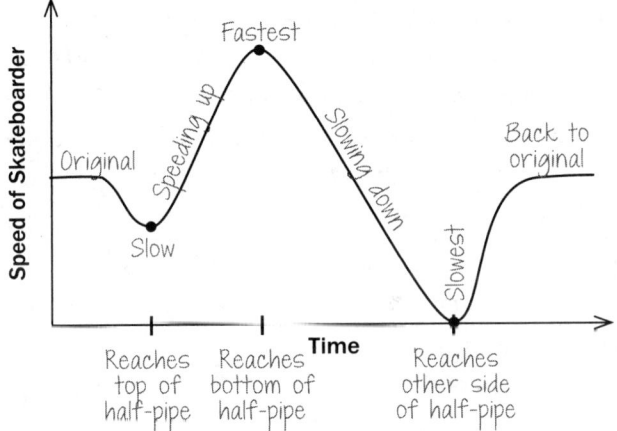

Finally, have the class talk through the story of the skateboarder one more time, but this time have two students at the board: one pointing out the story on the half-pipe diagram and one pointing out the story on the graph of speed over time.

To conclude, consider using one or more of these discussion questions:

» **What does the vertical dimension represent in the graph? In the picture?** In the graph, the vertical dimension is the speed of the skateboarder. In the picture, the vertical dimension shows the height of the skateboarder.

» **What does the horizontal dimension represent in the graph? In the picture?** In the graph, the horizontal dimension is time. In the picture, the horizontal dimension is horizontal position, the distance from the start.

? What if...

What if students think that the graph should look like the half-pipe picture?
Help students clarify what would go on the axes of such a graph (height and horizontal distance from the start). Discuss how that graph doesn't take speed or time into account (though they can be deduced). On a graph of speed versus time, the highest point (the maximum) will correspond to the highest speed (not height).

What if students insist on "lining up" the maximum and minimum points of their speed versus time graph with the corresponding locations on the picture of the skateboarder?
Students may be trying to graph speed versus horizontal distance. This kind of graph can give interesting information but is not the same as the graph of speed versus time. Discuss how the skateboarder travels more quickly on certain parts of the half-pipe, so it takes less time to cover certain distances—even some distances that are the same (such as from the left side of the half-pipe to the bottom and from the bottom to the right side).

» **Why doesn't the speed graph have the same shape as the picture of the skateboarder?** Listen for students who explain that the two images show the same story but different aspects of the same story. Have students describe the difference in their own words.

Student Problem Solving and Discussion

For **PROBLEMS 1–6**, students will need scissors, glue or tape, and the eight "Graphing Scenarios Cutouts" on page 49 of the Student Worktext. Have students work together to cut apart the scenarios and paste them beside the matching graphs. Note that this matching is not one-to-one. Students will match some graphs with two scenarios.

PROBLEMS 7–10 ask students to identify details in a scenario that might help them decide whether to "connect the dots" in a graph. Use the *Discuss & Write* box to have students share their ideas.

PROBLEMS 11–12 ask students to create and graph their own story to describe their sleepiness over a 24 hour period of time. Have them think about yesterday: when they were the most awake, when they were sleepiest, and, of course, when they were asleep.

Use these questions to discuss what to pay attention to in a graph and its related story:

» **The graph in problem 1 matches two scenarios. What phrases in the scenarios are similar to each other? How do these phrases show up in the graph?** Scenario B's "steady" temperature and scenario C's "flat" fee are represented by the horizontal section of the graph. In scenario B, the temperature "started rising," and in C, the plan costs "more for every minute." The graphs shows these ideas with a rising line (positive slope).

» **The graphs in problems 3 and 4 look very similar. How are the graphs different? How did you use that difference to figure out which scenarios match these graphs?** The graph in problem 3 starts at a positive value, whereas the graph in problem 4 starts at 0. Scenarios E and G match problem 3 because both scenarios have a non-zero initial value. Problem 4 matches scenario D because if you don't work any hours, you make $0.

» **Compare the graphs in problems 2 and 5. How are they similar? How are they different? What phrases in the scenarios helped you tell the difference between these graphs?** The graphs in problems 2 and 5 both show decreasing quantities, but the graph in problem 2 is curved and doesn't quite get to 0, while the graph in problem 5 does. Scenario A best matches the graph in problem 2 because "lots of candies get scooped up right away" and "a few are left." Scenario H goes with problem 5 because the cup "lets out a constant stream of coffee."

Students taking algebra may recognize the difference between graphs 3 and 4 as a difference in the *y*-intercept.

Lesson 6:
Solutions and Point Testing

PURPOSE

This lesson introduces students to the idea of a solution point. Students explore the idea that given an equation, *any* point on the coordinate plane either *is* or *is not* a solution point. An equation then serves as a "point-tester": any point can be put into the equation, and the equation will show whether the point is or is not on the graph of that equation. With this understanding of solution points, students develop a greater sense of what the graph of an equation represents: a graph is not just a picture; it is a representation of the *solutions* of the equation, the collection of *all* points that follow the pattern described by the equation.

The fact that for linear equations, the infinite collection of solution points happens to take the shape of a *line* when plotted together is special. This lesson helps students establish the relationship between solutions to an equation and points on a graph. They develop the idea that an equation represents a pattern of calculation that shows up as a visual pattern on the coordinate plane.

Understanding equations as point-testers can help students identify an equation by looking at a graph and make sense of non-linear graphs. The equation of a graph makes a true statement for *every* point on the graph and *only* for that collection of points. Writing an equation means describing a relationship between *x* and *y* for those points. This understanding extends to non-linear graphs. For students who understand a graph as a collection of an infinite number of points, the concept of a parabola extends from their understanding of a line; a parabola also marks out a specific collection of solution points to an equation on the coordinate plane.

Students will revisit these ideas in Unit 9: *Points, Slopes, and Lines* when they use point testing to generate linear equations using the knowledge that points on a line follow a pattern of constant rate of change and examine graphs of Inequalities where solution points encompass *regions* of the plane rather than a line or curve.

 **Mental Mathematics** *Begin each day with five minutes of Mental Mathematics (pages T51–T64). These activities help students build working memory so that they can keep multiple pieces of mathematical information in mind at once.*

Lesson at a Glance

Preparation

· Prepare to display the "Point Testing" grid on T42 for the Launch.

· *(optional)* The Launch activity suggests using four colors to mark points.

Mental Mathematics (5 min)

Launch: Point Testing (15 min)

· Place students in four groups, each with its own equation to use to test points.

· Discuss "point testing" as a way of investigating the graph of an equation; *any* point can be tested to see if it is a solution point, but once patterns start to emerge, making conjectures and testing specific points can further confirm a suspected pattern.

Student Problem Solving and Discussion (25 min)

· Give students time to work through the Important Stuff and additional problems.

· Ask students to share their responses from the *Discuss & Write* problem, which asks about the logic of connecting the solution points of an equation.

· Discuss what it means to be a solution point and what a graph really represents.

Unit 6 Related Game: Battleshape (See page T37 and Student Worktext page 43.)

Launch: Point Testing

Split the class into groups or let them work individually. Assign each group or individual one of the equations below:

- Group A will get the equation $y = x + 5$.
- Group B will get the equation $x + y = -1$.
- Group C will get the equation $2x + y = 2$.
- Group D will get the equation $y = 2x + 14$.

Display the "Point Testing" grid on page T42.

Circle the point (-3, 2) and ask students to test that point in their equation. "Testing" a point means using its x- and y-coordinates as values in the equation and seeing if the result is a true statement. Students in Groups A and B will find that (-3, 2) works for their equation. Have students show their evidence (e.g. 2 = -3 + 5 for Group A's equation). Assign each group a color and mark the point with the color(s) of the related equations.

Present the following six points one a time to the whole class, spending about a minute on each point for students to test the point and for the class to verify the results:

(-3, 2)	Satisfies equations A and B.
(-1, 4)	Satisfies equations A and C.
(-5, 4)	Satisfies equations B and D.
(3, -4)	Satisfies equations B and C.
(-3, 8)	Satisfies equations C and D.
(-9, -4)	Satisfies equations A and D.

Make sure students understand that this process could continue forever, because *any* point on the coordinate plane can be tested. As we test many points, we see that the solution points for a particular equation start to follow a visible pattern.

Mention that the four equations in this activity are special: not all equations have straight line graphs. But finding and testing solution points can always be used as a starting point for graphing equations. As students start to see a pattern emerging in the solution points for an equation, they can predict a solution point and test it. This is a way of gathering evidence to support or revise their hypothesis. Part of mathematics is then going further to understand why particular patterns emerge from particular equations.

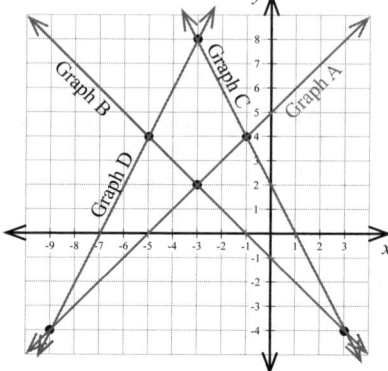

Here are some other ways to further the conversation:

- Ask students to predict another possible solution point for their equation and then to test it.
- Ask about the point (-7, -1). It *looks* like it may be on one of the lines, but point testing will confirm that it isn't.
- Ask about the points $(\frac{1}{2}, 5\frac{1}{2})$ and $(\frac{1}{2}, 5\frac{1}{4})$. Solution points don't have to have integer coordinates. Furthermore, *every* point is either on the graph (is a solution point) or it is not. Group A will find that their graph passes through $(\frac{1}{2}, 5\frac{1}{2})$ but not through $(\frac{1}{2}, 5\frac{1}{4})$. Students should get the message that graphing is not just about drawing a "straight line" and should understand that *every* point on that line is a solution for the equation.
- Ask students to predict and test other solution points (including points with non-integer coordinates) until they have found eight solution points for their equation. For these equations, but not for all equations, the graph is a line. Have students draw the line and choose one more point that is off the paper to test, with the equation, if it is on that line.

Student Problem Solving and Discussion

Allow time for students to work on the Important Stuff and additional problems.

Most of the problems are about testing or finding solution points. Help students see an equation as a statement that encodes a pattern of calculations. They have already approached equations in this way in the Think of a Number tricks from Units 1 and 5.

In this lesson, students start with an equation and use it to generate and test points. The repetition that students experience in this lesson is intentional and designed to build the idea that *every* point on a graph is related by the *same* pattern of calculation encoded in the equation. These points, taken as a collection, produce a graph.

Ask students to share written responses for **PROBLEM 8** in the *Discuss & Write* box. The problem is about whether it makes sense to "connect the dots" in the graph of an equation. Students have been thinking about when to connect the dots throughout Unit 6—most notably in Lessons 2 and 5. Listen for students who can explain that the line is just a way of indicating *every* point that follows the pattern of the solution points.

Use these prompts to support good thinking and discussion about graphs.

» **How do you know if a point is on the graph of an equation?** Students may think of an equation as a point-tester. For an equation that describes a relationship involving x and y, the x- and y-coordinates of any point can be plugged into the equation. The point is on the graph if the equation results in a true statement.

❓ What if...

What if students labor over each point in a sequence of problems?

Remind them that the calculations they are performing are actually the same each time because they are using the same equation. Help those students switch from focusing on "what to plug in where" to seeing $y = x + -5$ as an equation and thinking "in each point, y is five less than x" as a way to understand the pattern of calculation. It may help some students to think of this like a Mental Mathematics activity where they are given a prompt and have to subtract 5.

Make clear by your use of language that "solution points for an equation" and "points on the graph of the equation" are two names for the same set of points. The "graph of an equation" is not just what we see but a visual representation of the entire set of solution points for that equation. The "solution points" are not just the entries in the table but *all* the points that make the equation true. They therefore include any point that *could* be in the table.

» **Describe the relationship between the solution points of an equation and the graph of an equation. What about the non-solution points?** Listen for evidence indicating understanding that any point on the coordinate plane can be tested. Graphs of equations represent the set of solution points; furthermore, they include *all* solution points—even the ones that don't fit on the page. *Any* point that is *not* on the graph is a non-solution point, and any point that is not a solution point is *not* on the graph.

» **In problems 1–8, you worked with the equation $y = x + \text{-}5$. How does that compare with $x - y = 5$?** See if anyone can write the equations in a way that makes the comparison easier. Both can be converted, using legitimate algebraic steps, into the same form. By making the form of the equations as similar as possible, students can isolate specific features that might account for any differences between two graphs. In this case, the two equations represent the same relationship and so have the same solution points and therefore the same graph.

» **What if you were just given an equation (like $2x + y = 20$) and no possible solution points? What would you do to start figuring out what the graph looks like?** In this lesson, students were generally given solution points to test. But students should also realize that they can (and did) generate their own solution points. More of this will be addressed in Lesson 7. Students may start by testing arbitrary points, but they can be more strategic. They can think, for example, "What if x were 0?" and use this as an entry to generate solution points. Students should see that choosing "easy" values for x like 0, ±1, ±2, or ±3 (depending on the equation) is a useful strategy, but not the only correct approach. Later, they will learn that part of a strategy for graphing also includes recognizing features of an equation that make it possible to make predictions about whether the graph will be a line, parabola, circle, V-shape, or other shape.

Lesson 7:
Graphing Relationships

PURPOSE

This lesson helps students strengthen the connections among equations, tables, points, and graphs. Students generate their own table of solution points and explore the strategy of choosing "easy numbers" to substitute into the equation. Students must also make decisions about which solution points to examine and how many to find before being convinced that their graph, equation, and table all match. The matching activity gives students a sense of the overall shape of a graph and also gives students the option to use the graphs (and not just the equations) as "point-testers." Students should recognize how they can "plug" numbers into graphs to generate potential solution points that can be verified in an equation.

 **Mental Mathematics** *Begin each day with five minutes of Mental Mathematics (pages T51–T64). Students continue to build their ability to subtract, adjust, and keep track of their progress.*

Lesson at a Glance
Preparation
· Prepare to display "Graphing Equations" on page T43.
· Each student or group of students will need scissors and glue or tape for the "Matching Graphs to Equations" cutouts on page 51 of the Student Worktext.

Mental Mathematics (5 min)

Launch: Generating Solution Points (10 min)
· Students explore and discuss the strategy of using "easy numbers" to find solution points.

Student Problem Solving and Discussion (30 min)
· Give students time to work through the matching activity and explore additional problems.
· Discuss connections between equations, tables, points, and graphs.

Unit 6 Related Game: Battleshape (See page T37 and Student Worktext page 43.)

Launch: Generating Solution Points

Display "Graphing Equations" (page T43). Under the coordinate grid, write the equation: $x + 2y = 6$.

In the previous lesson, students either were given solution points to test or were asked to find solution points for specific values of x or y (when $x = 0$, for example). Now, students see that there is a table to fill out and a graph to draw, but they need to figure out how to start.

Suggest that *one* way of starting the problem might be to pick arbitrary points like (45, 62) or (-231, 6) to test. But finding solution points by substituting numbers that are "easy" to calculate with for one of the variables is more strategic. Emphasize this

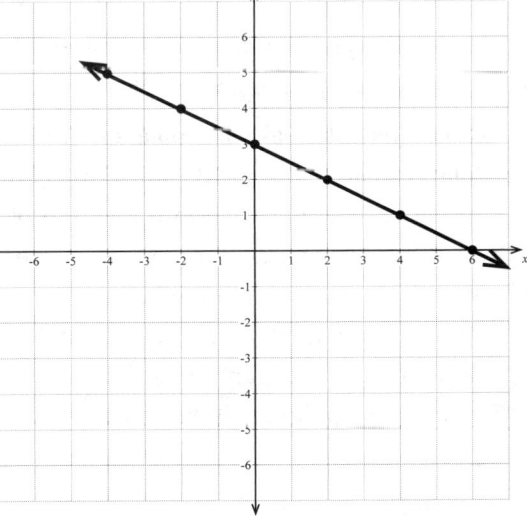

The *strategy* you are teaching is more important than the details of specific calculations. The focus of this initial activity is on choosing "easy numbers" to calculate with and graph.

Depending on what students have learned in their algebra class, it may also be useful to rearrange this equation into some other form, but that kind of thinking is not the focus of this lesson, and it certainly does serve students to be able to work with equations in a variety of forms.

Graphs, equations, and tables present different information about numerical relationships: tables give specific instances, equations describe the relationship precisely as a calculation, and graphs show the overall behavior. Part of mathematical proficiency is habitually asking oneself which tool presents the needed information in the clearest way. Students build flexibility with using equations to create tables, using tables to construct a graph, using equations to test points they think they see on a graph, and so on. Students use these different mathematical tools to help them make predictions or corroborate evidence about relationships and patterns they observe.

? What if . . .

What if students find no calculations "easy"?
It may help to use the idea of chunking from Unit 5. Once they see $x + 2y$ as a sum, they can think of number combinations that add to 6 and perhaps also involve an integer value for y (such as $4 + 2 \cdot 1$ and $-2 + 2 \cdot 4$). Also suggest that once they find several points that work, they may start to make educated guesses about where other solution points may be and then test those points.

strategy. "Easy numbers" are generally low (like 0, ±1, or ±2), but sometimes even low numbers give messy calculations. Fortunately (since *any* solution point will do), when a calculation gets complicated, it's fine to discard it and just try another.

Make sure students know that they get to choose a value for either x or y. Some students feel that they must plug in a value for x (only) and find y. This thinking has some merit—especially as a way to support the understanding of inputs and outputs of functions—but it isn't necessary. For example, in the equation $x + 2y = 6$, it is as convenient to choose a value for y as for x. Encouraging students to think flexibly about x and y will support their understanding of substitution in general.

Give students a couple of minutes to find some solution points for $x + 2y = 6$. Remind them they can discard messy calculations and that it might take some trial and error to find some "nice" points. For example, students may find that $x = 2$ leads to an easier calculation than $x = 1$.

Ask students to share their points and give a brief description of how they found them. For example, "I plugged in 0 for y and $6 + 0$ is 6, so my point is (6, 0)." Invite students to write one of their points in the table and graph it. Take four or five suggestions so students can see the line emerging. Leave at least one blank in the table.

Finally, ask students to use the emerging graph to predict solution points. Have them test these predictions in the equation, just as they did in Lesson 6. Fill in the remaining spaces in the table. Make sure students understand that the table doesn't always have to come before the graph; the graph can be used to fill in the table.

Student Problem Solving and Discussion

The Important Stuff is a matching activity. Students will need scissors, glue or tape, and a copy of the "Matching Graphs to Equations" cutouts on page 51 in the Student Worktext. Encourage students to work collaboratively.

The following questions ask students to think about connections between equations, tables, points, and graphs.

» **Share your strategies for filling out the tables. Did you make any observations that made the work easier?** Students may share ideas about using "easy" values. For example, in problem 7, x is multiplied by $-\frac{1}{2}$, so even-number values of x are easier to compute. Problems 6 and 8 involve x^2, which means that it is efficient to consider what happens for both positive and negative values of x. Students may share other strategies, such as looking at the graphs first, finding a point that is unique to a graph, and testing that point in all the equations.

» **Problem 2 asks about the equation $y = x + 4$. Problem 4 asks about the equation $y = 4$. Compare their tables and graphs. How are they similar? How are they different?** The graphs are both lines, and they both go through the point $(0, 4)$. This makes sense because if $x = 0$, then the y-values are the same $(y = 4)$. The x in the first equation, $y = x + 4$, makes a difference because whatever number you choose for x, that number gets added to 4. So it makes sense that the graph of $y = x + 4$ would climb higher as x takes on larger values, whereas the graph of $y = 4$ stays constant no matter what x is.

② Ⓒ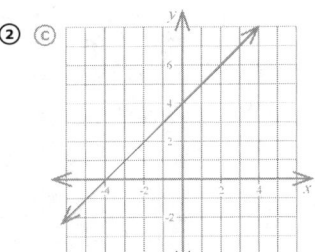

④ Ⓐ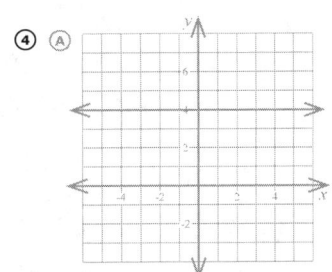

» **Problems 6 and 8 both produced graphs of *parabolas*. What are some observations you can share about parabolas?** Some students may have had prior experience with parabolas; others might be unfamiliar with the term and concept. The terminology is not important now. Instead, focus on what can be observed from the two problems. Students may notice that both of the equations involved a term with x^2. Ask students to consider what they know about squaring numbers. Students may mention the fact that squaring a positive number and squaring its opposite (negative) number give the same result. Help students connect such ideas to what they see in the graph.

⑥ Ⓓ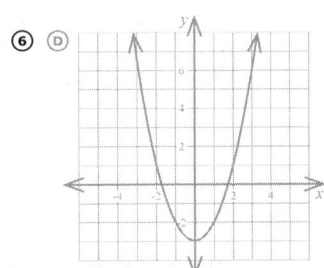

⑧ Ⓖ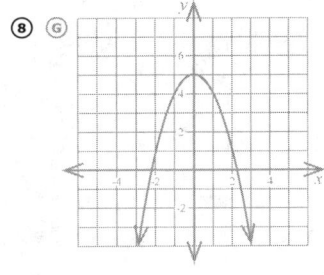

» **Describe different uses and advantages of an equation, a table of points, and a graph. How do you translate between them?** An equation describes a relationship between two numbers, x and y. It describes a pattern of calculation between them. A table of points lists some of the solution points for the equation. These are x and y pairs that make the equation true. For most equations, there are too many solution points to list, so a table of values lists only a few points. A graph is a picture of the solution points on one part of the plane. Every point on the graph could be a point on the table, and the coordinates of every point on the graph (just like the points in the table) make the equation true.

Student Reflections &
Unit Assessment

Before the Unit Assessment, ask students to reflect on the following:

What are some things you learned in this unit?

What questions do you still have?

Reflections can be done orally, on paper, or some combination of both. Use feedback from students to help them identify the big ideas from the unit and to select additional practice problems to help them prepare for the Unit Assessment included on pages T47 & T48. Before giving this assessment, consider spending a class period working through the Unit Additional Practice problems.

Since the Snapshot Check-in, students have:

- Explored graphs as stories and identified important features of a graph (such as increasing or decreasing sections, maximum or minimum points, where graphs cross the axes, etc.) in order to connect with information given in a narrative.
- Looked at graphs as representing infinitely many points, all of which make the corresponding equation true.
- Used equations as "point-testers," plugging a point into an equation to see if it produces either a true result (and is therefore a solution point) or a false result (and is therefore not a solution point).
- Seen that graphs and equations are two ways of showing the same relationship (graphs show a visual representation, while equations show a pattern of calculation that is true for every solution point).

Throughout Unit 6, students have focused on the following Algebraic Habits of Mind:

- **Communicating with Precision**—Students have seen that even without exact numbers, a graph can still communicate information about a story precisely through its direction (increasing, decreasing, or steady) and its shape (its steepness and whether it is straight or curved).
- **Using Tools Strategically**—Students have seen how a pattern expressed as an equation can be visually represented by a graph. They have seen how a graph itself can be used as a shortcut for calculations, since every solution point on a graph will make the associated equation true. And they have made connections among tables of values, graphs, and equations and have built facility in translating information between them.

→ EXPLORATION
Boxes of Chocolates

PURPOSE

Students seek and use structure in a pattern of chocolates and explore possible groupings that make it easier to find the total number of chocolates without counting each chocolate one by one. This kind of thinking illustrates "chunking," or looking for ways to organize complicated problems to make them simpler.

Students are then given a specific grouping structure for the problem. They extend this structure numerically (to larger boxes of chocolates), and then express the pattern algebraically. Thus, students also develop the algebraic habit of mind of describing repeated reasoning by extending their understanding of the structure of the problem to larger numbers and then algebra.

 **Mental Mathematics** *Begin each day with five minutes of Mental Mathematics (pages T51–T64). Establishing patterns of calculation helps students develop the habit of mind of seeking and using structure.*

Student Exploration and Discussion

PROBLEMS 1–4 explore patterns in the number of chocolates in a box as that box increases in size. Problem 1 asks students to think of a way to group the chocolates within each box so they are easier to count. If students are unclear, it may help to explain that the process is about finding a way to describe how the pattern grows. One way is by looking at the numbers, as in problems 3 and 4. But the pattern in the numbers is also connected to a visual pattern, and problems 1 and 2 are intended to help students see and express their pattern visually. Emphasize that the scheme or explanation students use should be applicable to larger and larger boxes of chocolates and that there are many ways to "see" the growth in the boxes of chocolates. The classroom discussion that follows should aim to capture this variety. Look for the different ways that students describe the visual pattern and consider how you might highlight the similarities and differences in a whole class discussion.

There are many possible grouping schemes to explore for problems 1–4. Here are three sample approaches you might see in your classroom.

Possible Grouping Scheme 1 (presented in Answer Key)

The diagram below groups the box into columns containing increasing consecutive odd numbers of chocolates up till the middle row, after which the columns contain decreasing consecutive odd numbers. In any box b,

Exploration at a Glance

Preparation *(optional)*
- Students may find it helpful to work with manipulatives like coins, counting chips, or candies.
- Colored pencils might be useful for students to represent their grouping strategies.

Mental Mathematics (5 min)

Student Exploration and Discussion (40 min)
- Provide time for students to explore problems 1–4 on their own or in small groups.
- Help students focus on the process of grouping the chocolates to simplify counting; this is the key mathematical idea in this activity.
- Have students present their grouping schemes from problems 1–4, and discuss connections between them.
- Allow students to explore problems 5 & 6 on their own or in small groups.
- Discuss how the expression found in problem 5 relates back to the chocolates and how different grouping schemes (or "chunks") help to show the structure of a problem.

Unit 6 Related Game: Battleshape (See page T37 and Student Worktext page 43.)

Algebraic Habits of Mind

Seeking and Using Structure
In problems 1–4, students look for a possible grouping they can use to help describe how the pattern in the boxes of chocolates grows. This possible grouping can take many forms, but they are all a part of seeking structure in the problem.

the middle column has $2b + 1$ chocolates (one in the center, b above, and b below). So, to find the number of chocolates in Box #8, we must add the odd numbers from 1 to 17, then back down from 15 to 1 (that is, $1 + 3 + \ldots + 15 + 17 + 15 + \ldots + 3 + 1$).

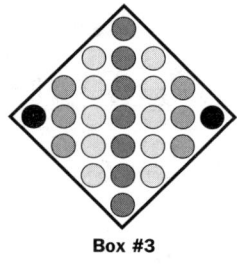

Box #3

Another way to do this calculation would be to find 17 plus twice the sum from 1 to 15 ($2(1 + 3 + \ldots + 15) + 17$). This grouping scheme shows how the number of chocolates grows between consecutive boxes. For example, Box #8 has $15 + 17 = 32$ more chocolates than Box #7 (which has 15 in the middle column), and more generally, box b has $(2b - 1) + (2b + 1) = 4b$ more chocolates than the previous box.

Possible Grouping Scheme 2

This scheme interprets the growth as adding an "outer layer" of chocolates to the number in the previous box (shown in black in the diagrams). Box #1 has 4 chocolates in the outermost layer, Box #2 has 8, Box #3 has 12, and, more generally, box b contains $4b$ chocolates in the outermost layer.

These two sets of diagrams show, in two different ways, that the outer layer of chocolates is a multiple of 4. In the top set, the organization scheme shows how each side of the outer square layer gets larger by one extra chocolate. The bottom set shows another way to see that the outer layer always contains a multiple of 4; again each of the four sides of the square grows by one additional chocolate.

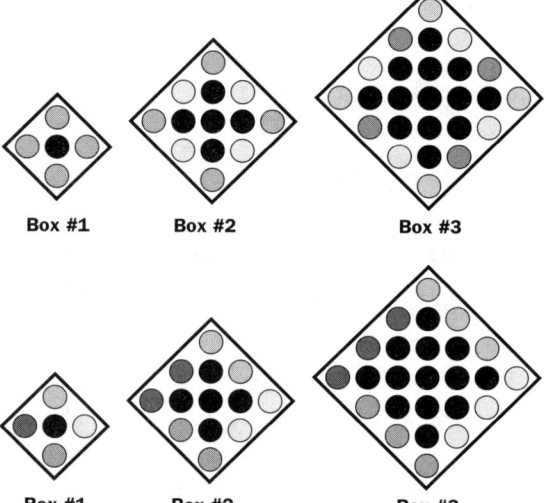

Box #1   **Box #2**   **Box #3**

Box #1   **Box #2**   **Box #3**

This grouping scheme shows why the number of chocolates in every box is one more than a multiple of 4: there is one in the center, surrounded by layers that contain multiples of 4. This also helps us see that in general, box b contains $4b$ more chocolates than the previous box. For example, Box #8 will have $4(8) = 32$ more chocolates than Box #7. Finding the total number of chocolates in any box is a little more tricky. For example, Box #3 contains $1 + 4(1) + 4(2) + 4(3)$ chocolates. Written another way, this is $1 + 4(1 + 2 + 3) = 25$ chocolates. By extension, Box #8 will contain $1 + 4(1 + 2 + \ldots + 8)$ chocolates. Generally, box b will contain $1 + 4(1 + 2 + \ldots + b)$ chocolates. In future Explorations, students will become much more familiar with the sum $1 + 2 + 3 + \ldots + n$ (a sequence called the "triangular numbers") and will discover a way to find this sum for any number n.

Possible Grouping Scheme 3

This grouping scheme shows a different way to analyze growth by keeping track of what's already been counted and then seeing how the extra chocolates are added. This time, the chocolates from the previous box are counted as a square in the left corner of the current box. The organizational scheme shows again how the growth in the number of chocolates can be seen in multiples of four. Again, box b contains $4b$ more chocolates than the previous box.

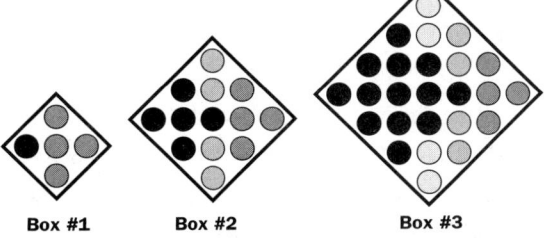

Box #1   Box #2   Box #3

Students may come up with yet more ways to group the chocolates. Some may even anticipate the schemes shown later in problems 5 and 7. Encourage alternate approaches and use them to focus your class discussion.

In **PROBLEM 5**, students explore a grouping scheme of dark and white chocolates that produces a familiar set of numbers (perfect squares) and work toward an algebraic expression. Communicate that there is no such thing as a "best" scheme, but that different schemes suggest different insights. The scheme in problem 5 leads to one convenient way of expressing the number of chocolates in any size box b. Students should understand that the formula $b^2 + (b + 1)^2$ is a way of expressing "the sum of two consecutive square numbers" and be able to describe how this formula is directly related to the visual pattern of the white and dark chocolates.

Connect the picture of the dark and white chocolates to the boxes explored in problems 1–4. In particular, make sure that students see that even though the expression works to get "answers" for the boxes of chocolates in problems 1–4, it still requires *explanation* to show how to link the expression with the visual grouping pattern that students observed. This may be as simple as explaining how the chocolates in the boxes in problems 1–4 could be "chunked" into two squares like in problem 5.

Here are some more ways to look at the structure of this problem and connect it to what students have found algebraically.

> » **Square numbers can be written as the sum of consecutive odd numbers:** $n^2 = 1 + 3 + 5 + \ldots + (2n - 1)$. **Find a way to connect the pictures of boxes of chocolates to this relationship between square numbers and odd numbers.** Students examined this idea in the Exploration "Building Squares" in Unit 4: *Area and Multiplication*. With the boxes of chocolates, students can show how, for example, $1^2 + 2^2 = (1) + (3 + 1)$. And $2^2 + 3^2 = (1 + 3) + (5 + 3 + 1)$. Ask students to demonstrate different ways they see this connection. Some might see the connection more easily with the numerical representation. Others might have a clear way of

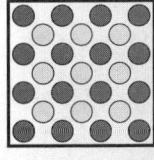

showing the connection between the pictures of the milk chocolates and the picture of the white/dark chocolates by showing, for example, that $1 + 3 + 1$ represents the number of milk chocolates in each vertical column in Box #1, while $(1) + (1 + 3)$ represents the number of (white) + (dark) chocolates. The algebraic habit of mind of seeking and using structure includes being able to connect different representations of a problem and move fluidly among them.

» **Describe how you can use the algebraic expression from problem 5, $b^2 + (b + 1)^2$, to show why the difference between the numbers of chocolates in two consecutive boxes is always a multiple of 4.** The total number of chocolates in box b is $b^2 + (b + 1)^2$. The total number of chocolates in the next larger box is $(b + 1)^2 + (b + 2)^2$. If we want to know how many more chocolates are in the larger box, we can subtract the two quantities: $((b + 1)^2 + (b + 2)^2) - (b^2 + (b + 1)^2)$. This is a complicated expression, but encourage students to see the larger structure of this expression. Invite them to see that this is the same as $(b + 1)^2 + (b + 2)^2 - b^2 - (b + 1)^2$. And no matter how complicated a chunk it is, if we see the *same* chunk being added and subtracted, then we can remove them both. So the expression can be simpler: $(b + 2)^2 - b^2$. Now, students can use an area model to rewrite the expression as $b^2 + 4b + 4 - b^2$. This simplifies to $4b + 4$, which is a multiple of 4. Help students make visual sense of this algebraic process and connect it to observations about a pattern in the "outermost layer" of chocolates from problems 1–2.

$(b + 1)^2 + (b + 2)^2$ $b^2 + (b + 1)^2$ $4b + 4$

Further Exploration

PROBLEM 7 presents yet another grouping scheme for the chocolates, and PROBLEM 8 asks students to connect the algebraic expression from problem 7 with the algebraic expression they found in problem 5.

This is considered "Further Exploration" because the main part of the Exploration activity is in problems 1–4, when students work to establish their own organizational scheme for tackling the problem. The Further Exploration problem is included to show that the grouping shown in problem 5 is not the only way to generate an algebraic expression. This alternate way of grouping the chocolates shown in problem 7 leads to a different but equivalent algebraic expression, $2b^2 + 2b + 1$.

Algebraic Habits of Mind

Describing Repeated Reasoning

After exploring several numerical examples, students create equations to record their process of calculation. These equations can be used to make generalizations about all boxes or, in this case, relationships between any two consecutive boxes. Remind students that they can always fall back on a numerical example to help them understand what is going on algebraically. Students who describe repeated reasoning may use many numerical examples and return to them as they consider expressions; they may also use English as well as algebra to describe the patterns they see but will eventually repeat their calculation with a variable to record and generalize the process.

→ EXPLORATION
Street Paths

PURPOSE

In this Exploration, students draw paths along gridlines on the coordinate plane and measure these paths between specified points. As students count the number of shortest paths from a specified starting point to points one and two blocks north of it, they encounter familiar numerical sequences: the counting numbers and the triangular numbers. Students have seen these sequences emerge in several contexts and are starting to develop the habit of going beyond simply identifying a pattern to looking for ways to explain how the pattern arises.

 *Mental Mathematics* Begin each day with five minutes of Mental Mathematics (pages T51–T64). These lively activities help build students' confidence, competence, and mental capacity.

Launch: Taxicab Geometry

Explain to students that in this Exploration, they can travel only along gridlines (they can't go through squares diagonally). They are looking for the shortest path from point A to point B, discovering how many routes achieve this shortest length.

There is a whole area of mathematics dedicated to "taxicab geometry," or how geometry is affected by staying on gridlines. In exploring taxicab geometry, students experience seeing familiar ideas in a new and different way. This experience can help students think more flexibly and with more depth. For example, a student may realize that measuring distance along gridlines gives results that "feel" wrong. In taxicab geometry, both of these graphs show points that are the same distance (4 units) apart; yet, the points in the second graph "feel" closer than the first, and of course, in the usual geometry of the coordinate plane, they are.

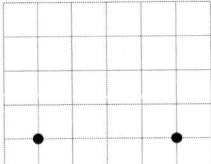 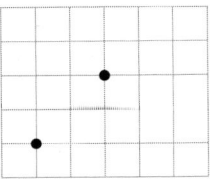

This thinking may lead a student to recognize that in order to find the distance between two points on a coordinate plane, we need a different system. Finding distance using the Pythagorean equation ($a^2 + b^2 = c^2$) will be explored in Unit 9: *Points, Slopes, and Lines.*

Exploration at a Glance

Preparation: *(optional)* Students might find it helpful to have extra graph paper.

Mental Mathematics (5 min)

Launch: Taxicab Geometry (5 min)

· Explain that students can travel only on gridlines (not diagonally) and that they are comparing the lengths of different paths along these gridlines.

Student Exploration and Discussion (35 min)

· Provide time for students to explore on their own or in small groups.

· Discuss and extend problem 5, which asks students to generalize their observations, and then challenge students to consider other similar questions.

· Even if students don't have a chance to go through all of the Further Exploration problems, have them discuss the numerical patterns from problems 6 and 7. Encourage students to make conjectures about *why* the patterns arise as they do.

Unit 6 Related Game: Battleshape (See page T37 and Student Worktext page 43.)

Some familiar ideas like "distance" or "circle" look different in taxicab geometry. This mathematics is useful when thinking about maps; for example, students may think about algorithms used to determine the best route in navigation systems, especially considering how many possible shortest paths there may be (as students will see in the Exploration).

Student Exploration and Discussion

In PROBLEMS 2–5, students look at two specific points, A and B. They examine the possible shortest paths, then the next shortest paths, and then the next shortest paths.

In problem 5, students make the conjecture that every path length between A and B is an odd number of blocks. Discuss this idea as a class. Change the locations of A and B and ask students to rethink the questions in problems 2–5. Students may think that path lengths between *any* two points must be odd. There are many counterexamples. Students should also start to see how to describe a method for finding a shortest path between any two points. For example, if B is 3 blocks north and 4 blocks east of A, a student might reason: "Start at A. Every move must be to the north or east; going south or west makes the path longer. We must move 3 blocks north and 4 blocks east, but the order doesn't matter. *Every* path between A and B that goes exactly 7 blocks will be a shortest path. Any other path will be longer."

To go further with the idea, ask students to explain why it makes sense that if the shortest path between two points is an odd number of blocks, then *any* path between those two points will also be an odd number of blocks. Students might come up with an explanation like "If you're not on the shortest path, then you went at least one block extra in some direction and you'll have to move back one block to make up for it. So you're adding an even number of blocks."

Students may also discuss patterns in the tables from PROBLEMS 6 & 7 (see Further Exploration). If students don't have time to go through the Further Exploration, consider using those ideas as part of a whole class discussion.

Further Exploration

In PROBLEM 8, students examine the pattern they identified in problem 6 by "listening in" on conversations between characters describing their systematic approach to the problem. The characters demonstrate mathematical thinking in their descriptions and when they look for "shortcuts" once they recognize repeated reasoning. For example, Ben and Carla both notice that they repeat most of their process as they look for the shortest paths for points further and further from the initial point **L**.

PROBLEM 9 asks students to extend this thinking to examine the more complicated pattern identified in problem 7. These problems lead students to examine their process as they make sense of the pattern. In problem 7, students see the triangular numbers (1, 3, 6, 10, 15, etc.), a pattern that shows up in many contexts and that students might start to recognize. Future Explorations will bring this number sequence back and help students work with it.

Game:
Battleshape

PURPOSE

This game gives students an enjoyable way to practice naming coordinate pairs in the context of the geometry of shapes on the coordinate plane (which they also explore using transformations in Lessons 3 and 4). It previews the idea of point testing (used in Lesson 6) to identify points as either solution points or non-solution points and encourages deductive reasoning and communicating with precision as students must use limited information to re-create their opponent's graph.

Game Suggestions

- Game instructions are on page 43 of the Student Worktext.

- On the first day, demonstrate game play. If students are mostly unfamiliar with the game format, display the game board and play a round against a small group of students who are collaborating at their own game board hidden from your view.

  - Demonstrate drawing one of each of the five shapes on the top half of your game board, and ask the student players to draw the same five shapes, one of each, on their own board, arranged however they like, but not overlapping.

  - Demonstrate playing against the group. Guess a point, and instruct them to say "hit" or "miss" and, if it's a hit, to say which shape it is. Show how to indicate that result on the lower half of the displayed game board, circling the coordinate point to show a hit or marking it with an X for a miss. Then ask the group to guess a point. For this demonstration purpose, it's okay that the students see your displayed board and "cheat." Go through several turns.

  - You may also wish to demonstrate recording the shapes "hit" by listing one of the coordinates together with the name of the shape along the side of the page. This can help students keep track of information they've found and encourages good record keeping.

- Students will likely get a chance to play one or two games in 5 to 10 minutes.

Variations

- Vary the shapes used. For example, one day, students might play the game using only rectangles—they can draw three rectangles of *any* size on their board, and when their opponent hits one of their rectangles, they have to indicate the size of the rectangle using its dimensions (e.g., "You hit a 2×5 rectangle.").

- Vary the game board. Have students create their own grids using graph paper. Students should agree on minimum and maximum values for the horizontal and vertical axes so that they use the same playing space.

- Play with teams of two against two. Students are more likely to talk about strategy—both for avoiding being hit and for finding shapes. This will likely be more productive after students have had a chance to play the game individually a few times to get the hang of the rules.

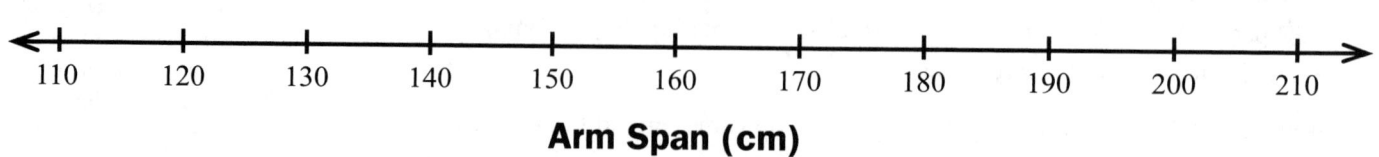

Arm Span (cm)

Coordinating Arm Span and Height on the Coordinate Plane (Lesson 1)

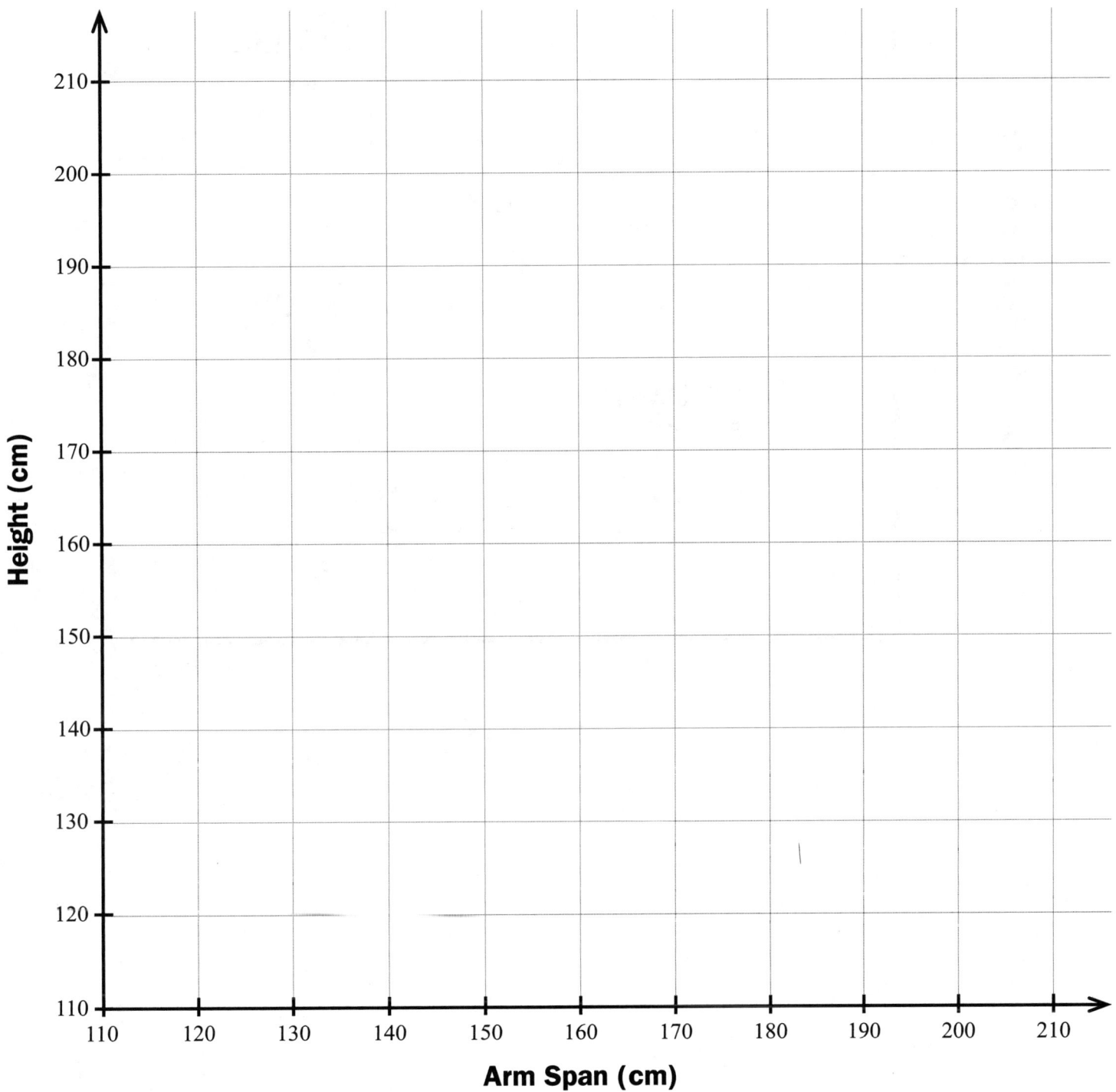

Arm Span (cm)

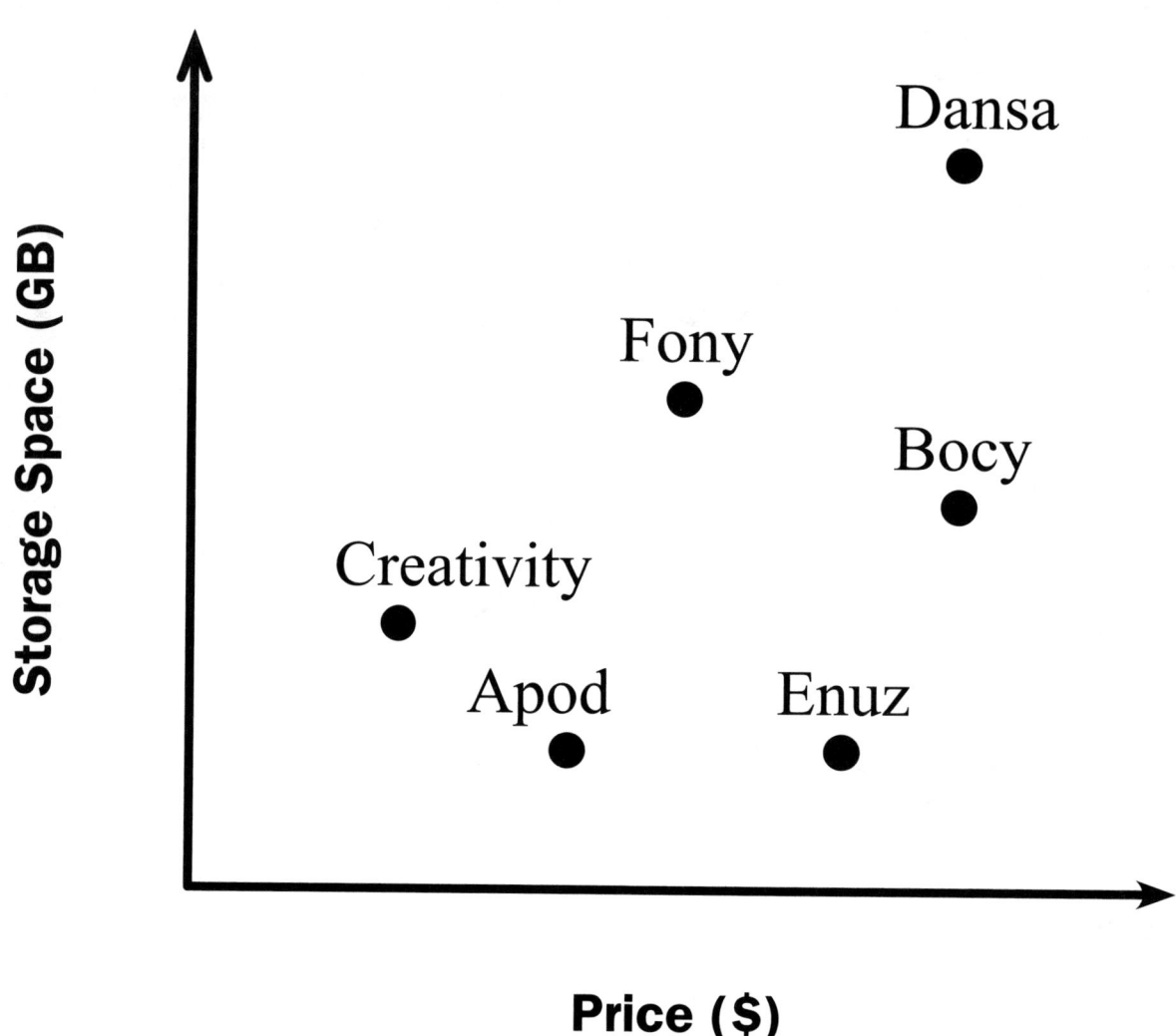

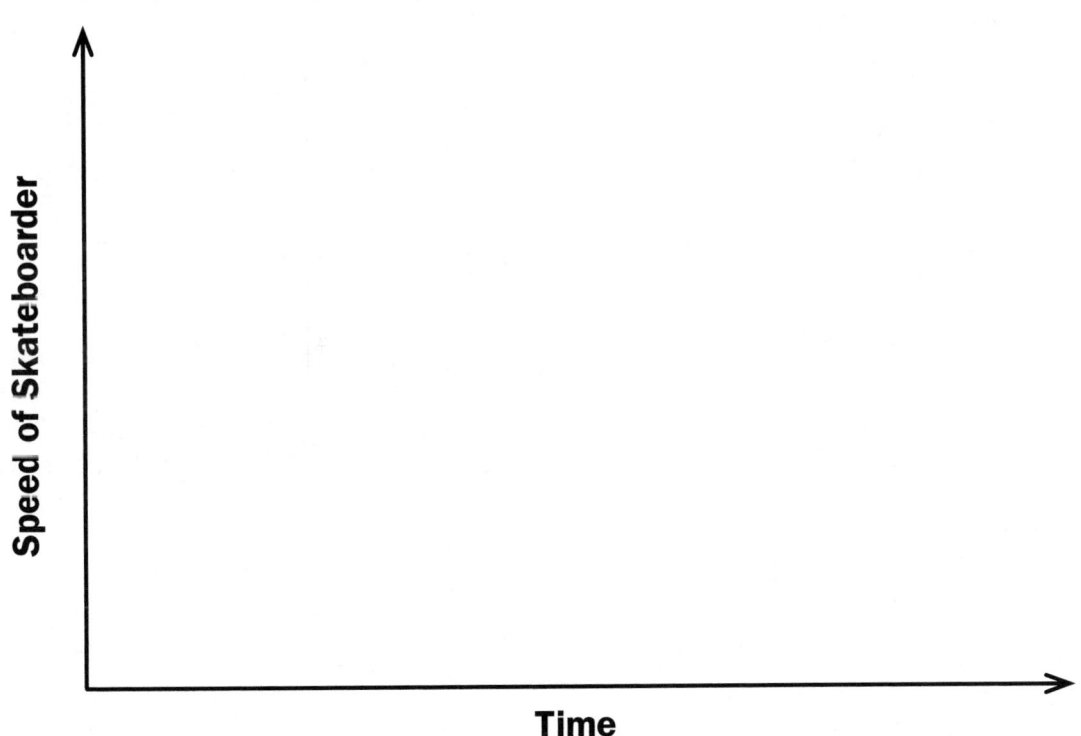

Time

Speed of Skateboarder

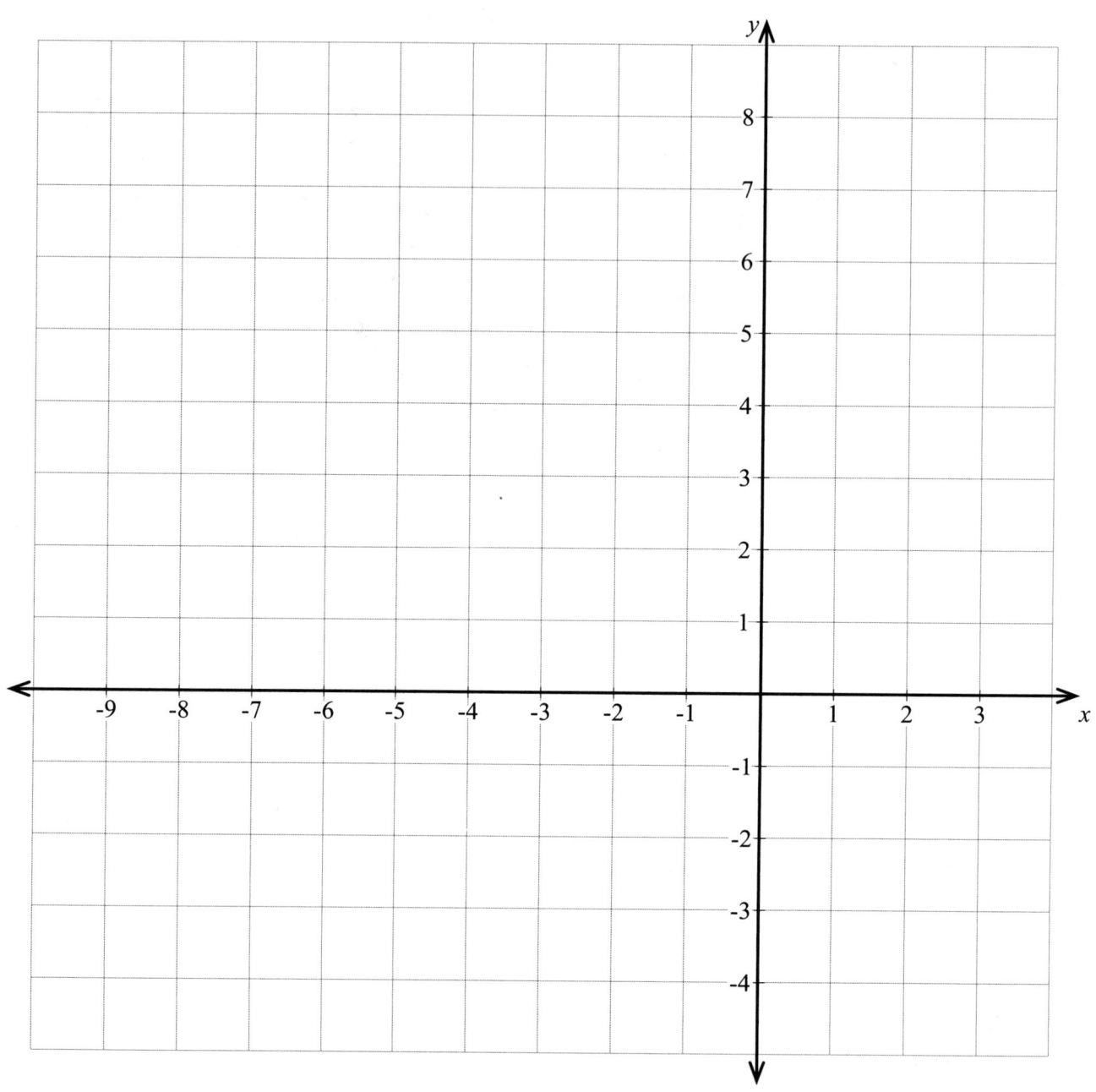

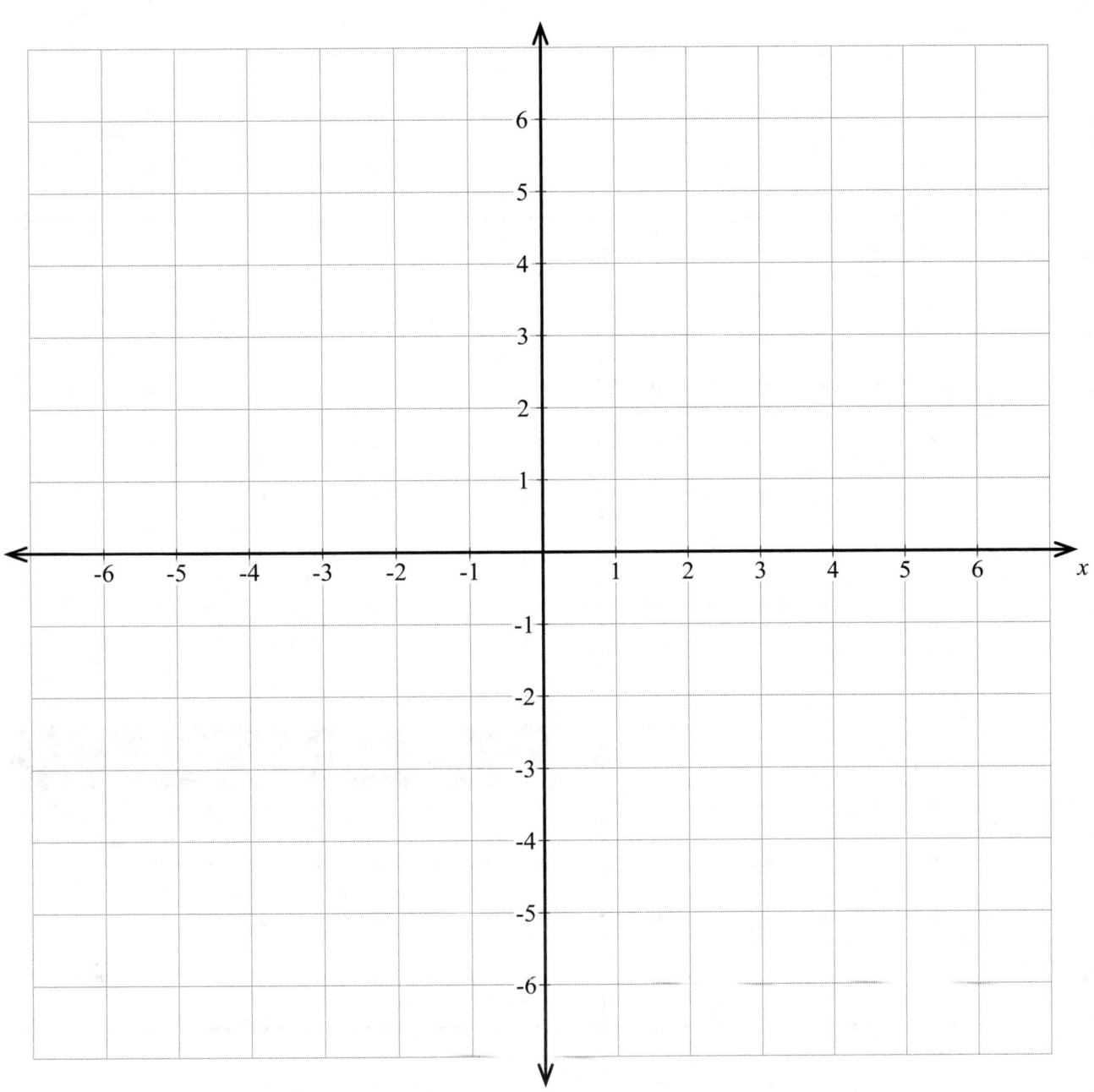

Equation:

x							
y							

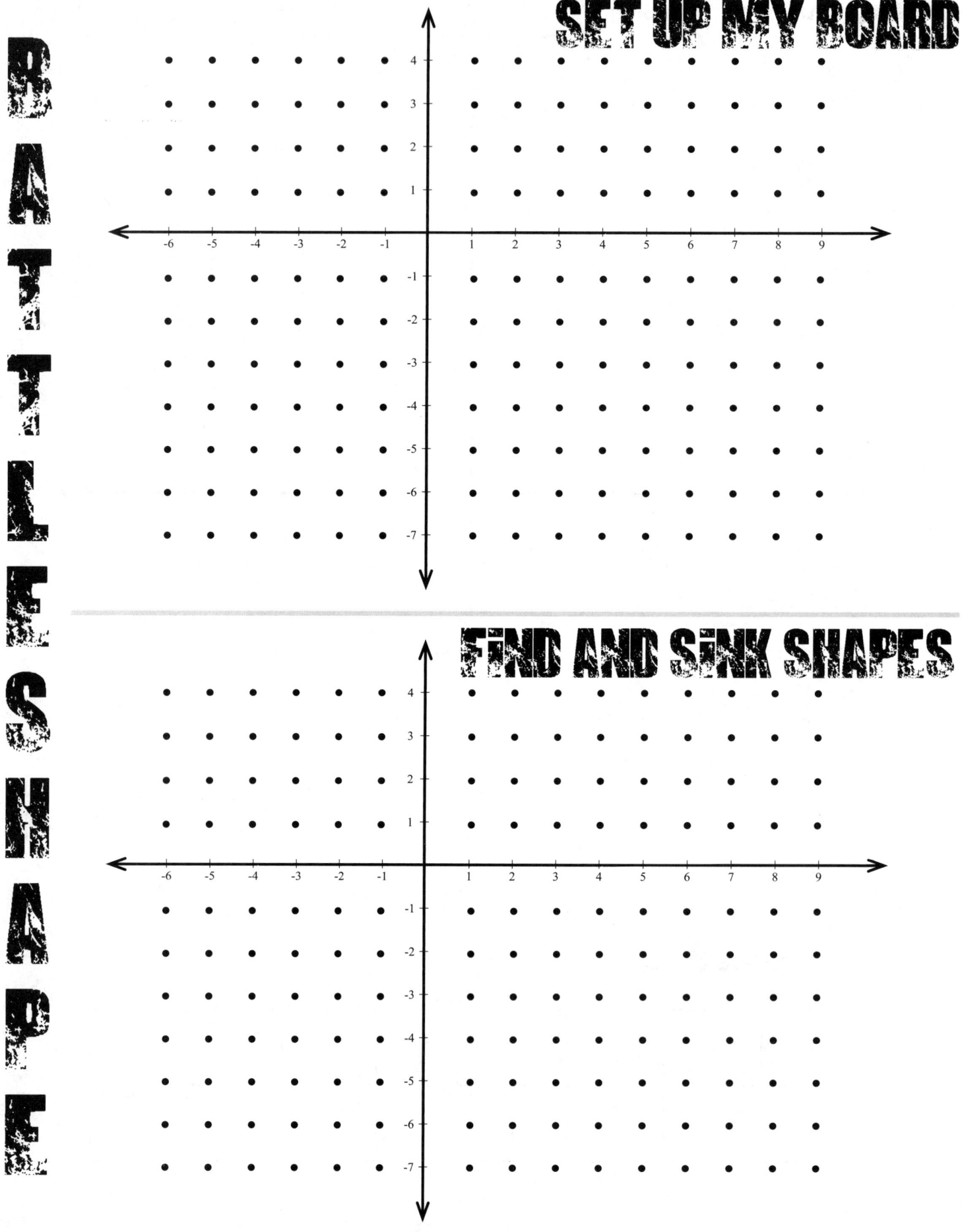

SET UP MY BOARD

FIND AND SINK SHAPES

Snapshot Check-in

Name:

Fill in the coordinates and draw and label each point.

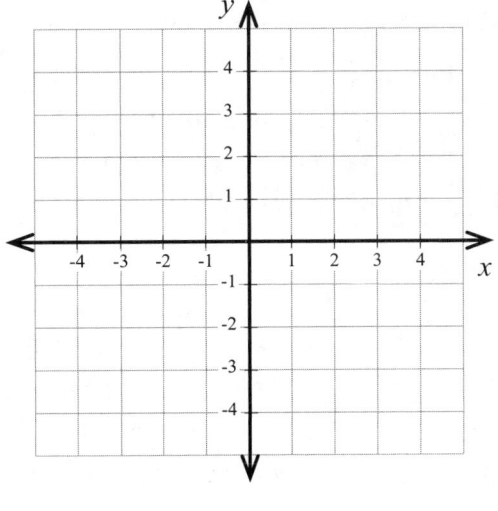

(1) I am point A.
- My *x*-coordinate is 3.
- My *y*-coordinate is -4.

Where Am I? (,   )

(2) I am point B.
- My *x*-coordinate is -2.
- *y* = *x*

Where Am I? (,   )

(3) I am point C.
- My *x*-coordinate is 2.
- My *y*-coordinate is one more than my *x*-coordinate ($y = x + 1$).

Where Am I? (,   )

(4) Ian, Jacob, Kayla, Luis, and Mali measure their heights and arm spans. They graph the results.

(a) Who is the tallest?

(b) Which two students have the same arm span?

(c) Compare Kayla and Ian. Who is taller? Who has longer arms?

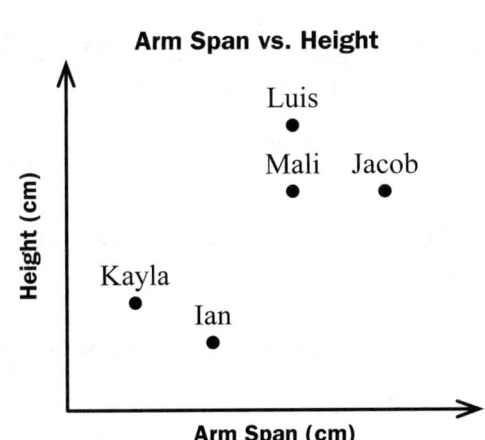

Arm Span vs. Height

(5) Use the transformation rule $(x, y) \rightarrow (x + 5, y - 4)$ to move and redraw the graph in the space below. Record points in the table.

(x, y)	(-1, 5)	(1, 4)	(1, 2)		
$(x + 5, y - 4)$				(4, -2)	(3, 0)

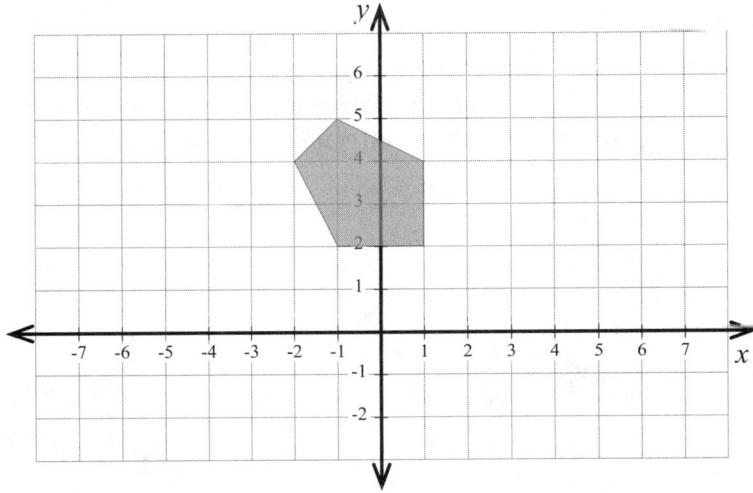

(6) Use the $(x, y) \rightarrow ($, $)$ notation to show the transformation rule for a shape that moves left 3 units and up 8 units.

Snapshot Check-in

Fill in the coordinates and draw and label each point.

① I am point A.
- My *x*-coordinate is 3.
- My *y*-coordinate is -4.

Where Am I?   (3 , -4)

② I am point B.
- My *x*-coordinate is -2.
- *y = x*

Where Am I?   (-2 , -2)

③ I am point C.
- My *x*-coordinate is 2.
- My *y*-coordinate is one more than my *x*-coordinate ($y = x + 1$).

Where Am I?   (2 , 3)

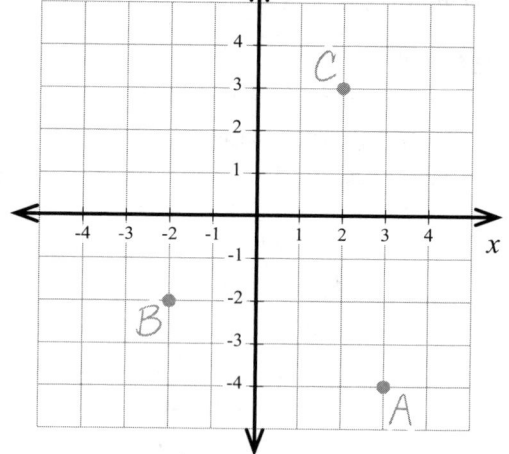

④ Ian, Jacob, Kayla, Luis, and Mali measure their heights and arm spans. They graph the results.

(a) Who is the tallest?

Luis

(b) Which two students have the same arm span?

Luis and Mali

(c) Compare Kayla and Ian. Who is taller? Who has longer arms?

Kayla is taller. Ian has longer arms.

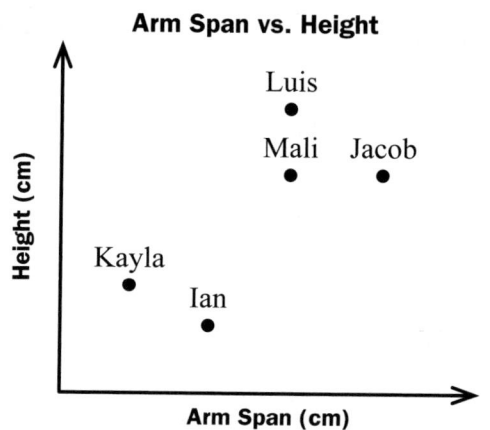

Arm Span vs. Height

⑤ Use the transformation rule $(x, y) \rightarrow (x + 5, y - 4)$ to move and redraw the graph in the space below. Record points in the table.

(x, y)	(-1, 5)	(1, 4)	(1, 2)	(-1, 2)	(-2, 4)
(x + 5, y − 4)	(4, 1)	(6, 0)	(6, -2)	(4, -2)	(3, 0)

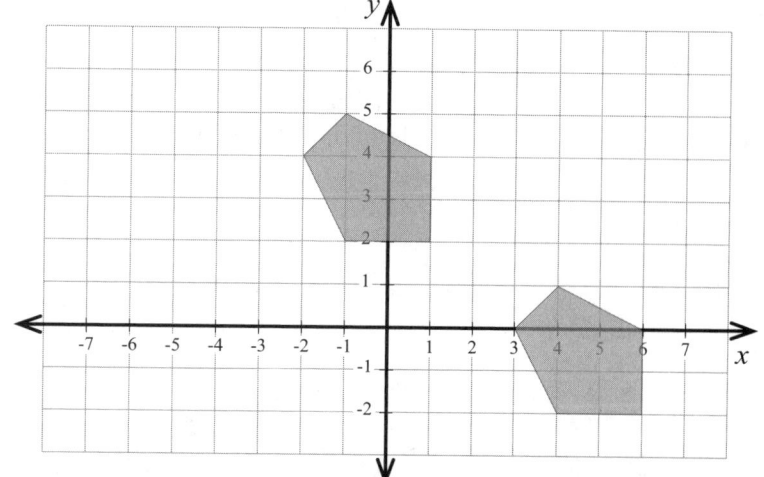

⑥ Use the $(x, y) \rightarrow (\ , \ )$ notation to show the transformation rule for a shape that moves left 3 units and up 8 units.

$(x, y) \rightarrow (x - 3, y + 8)$

Unit Assessment

Find the coordinates. Plot and label each point.

① I am point A.
- My x-coordinate is 1.
- My y-coordinate is -3.

Where Am I? (,   )

② I am point B.
- $x + y = -7$
- My x-coordinate is -4.

Where Am I? (,   )

③ I am point C.
- My x-coordinate is twice my y-coordinate ($x = 2y$).
- My vertical position is 2.

Where Am I? (,   )

④ I am point D.
- Start at A and use the rule $(x, y) \rightarrow (x - 4, y + 6)$.
- $y = -x$

Where Am I? (,   )

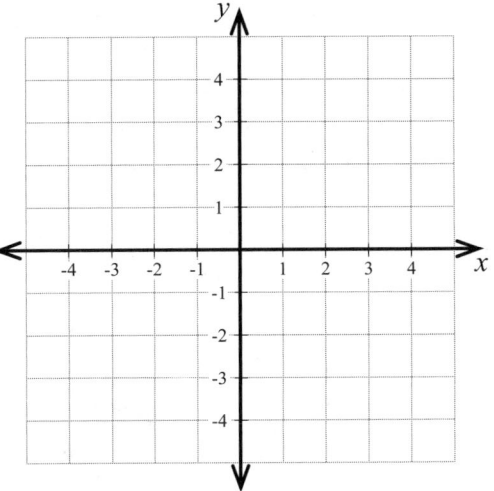

Find five solution points for each equation. Match each equation to a graph. (One graph will not be needed.)

⑤ $x + y = -3$ Graph: _____

x				
y				

⑥ $2x + 1 = y$ Graph: _____

x				
y				

Ⓐ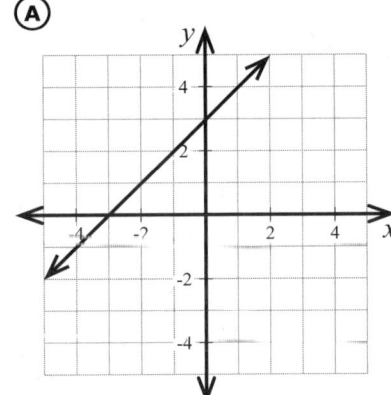

Ⓑ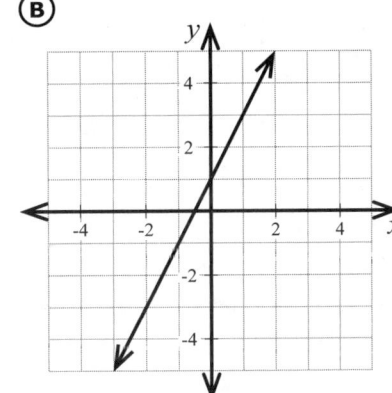

Ⓒ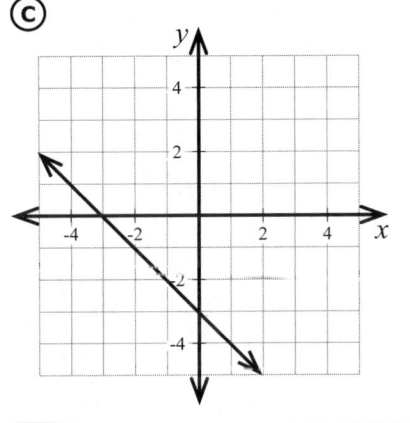

---

⑦ Use the transformation rule $(x, y) \rightarrow (2x, 3y)$ to fill out the table and redraw the shape.

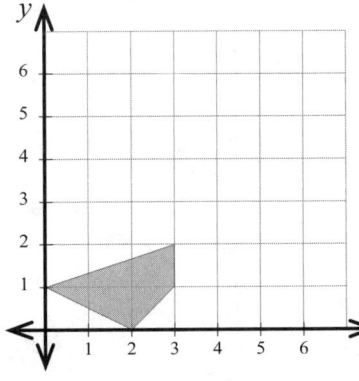

(x, y)	$(2x, 3y)$
$(0, 1)$	
$(3, 2)$	
$(3, 1)$	
$(2, 0)$	

⑧ Use the transformation rule $(x, y) \rightarrow (x + 6, y - 2)$ to move and redraw this graph. Record the points in the table.

(x, y)	$(-2, 4)$	$(1, 3)$		$(-1, 0)$		
$(x + 6, y - 2)$			$(6, -1)$		$(4, -2)$	$(3, 0)$

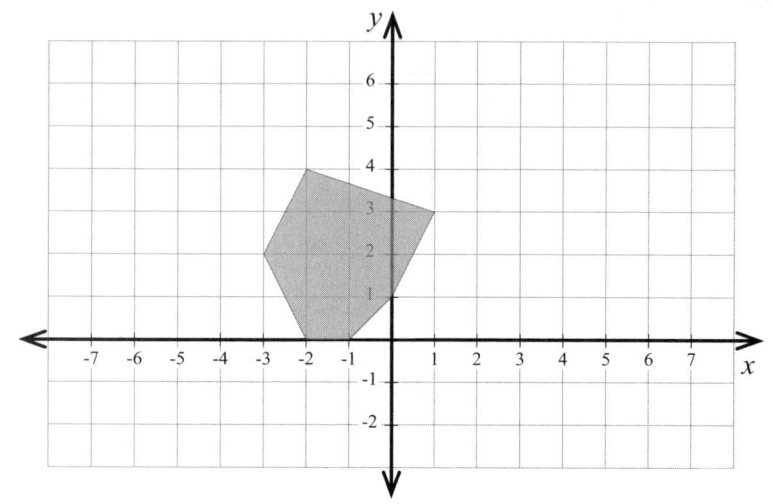

⑨ Use the $(x, y) \rightarrow (\quad , \quad)$ notation to show the transformation rule for a shape that moves right 4 units and up 1 unit.

⑩ Kayla's body temperature had been normal all morning, but it began rising steadily after noon. Which graph shows her temperature during the day?

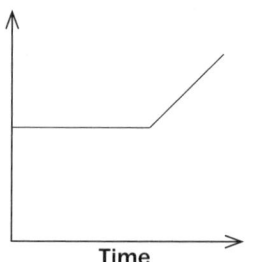

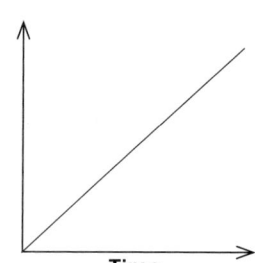

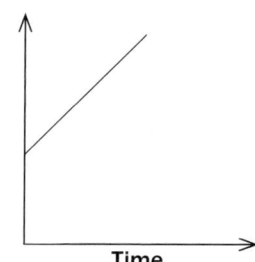

⑪ Find solution points for $y = 1 - x$.

a Find a solution point if $x = 0$.

b Find a solution point if $x = 2$.

c Find a solution point if $x = -3$.

d Find a solution point if $x = 20$.

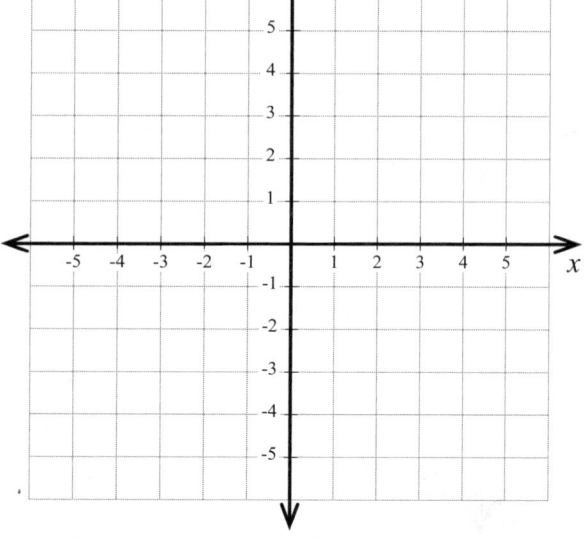

e Circle the solution points and cross out non-solution points.

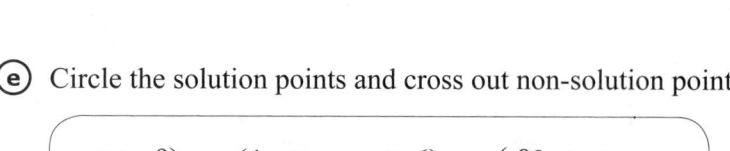

$(10, -9) \quad (4, -3) \quad (5, 6) \quad (-99, 100)$

f Plot the solution points that fit on the graph paper and use them to draw the graph of $y = 1 - x$.

Unit Assessment

Find the coordinates. Plot and label each point.

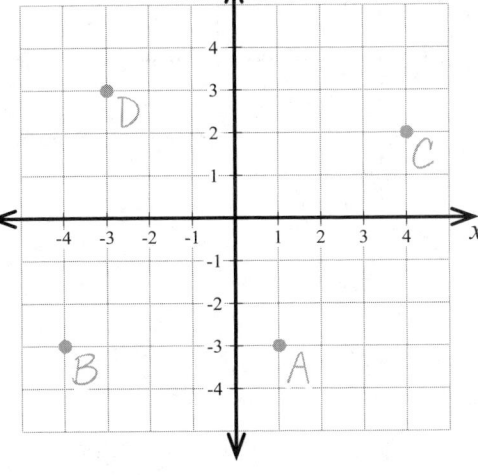

① I am point A.
- My x-coordinate is 1.
- My y-coordinate is -3.

Where Am I?   (1 , -3)

② I am point B.
- $x + y = -7$
- My x-coordinate is -4.

Where Am I?   (-4 , -3)

③ I am point C.
- My x-coordinate is twice my y-coordinate ($x = 2y$).
- My vertical position is 2.

Where Am I?   (4 , 2)

④ I am point D.
- Start at A and use the rule $(x, y) \rightarrow (x - 4, y + 6)$.
- $y = -x$

Where Am I?   (-3 , 3)

Find five solution points for each equation. Match each equation to a graph. (One graph will not be needed.)

⑤ $x + y = -3$ Graph: __C__

x	-1	0	1	(Responses
y	-2	-3	-4	will vary.)

⑥ $2x + 1 = y$ Graph: __B__

x	-1	0	1	(Responses
y	-1	1	3	will vary.)

Ⓐ No match

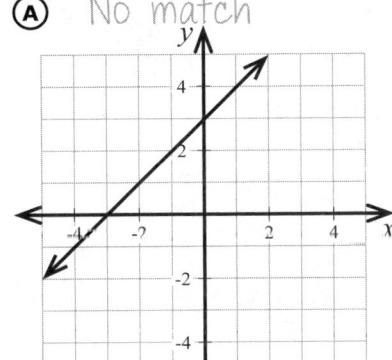

Ⓑ Equation 6

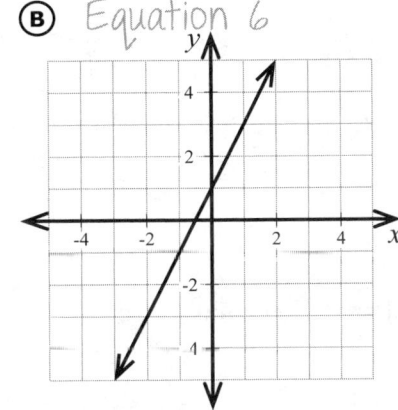

Ⓒ Equation 5

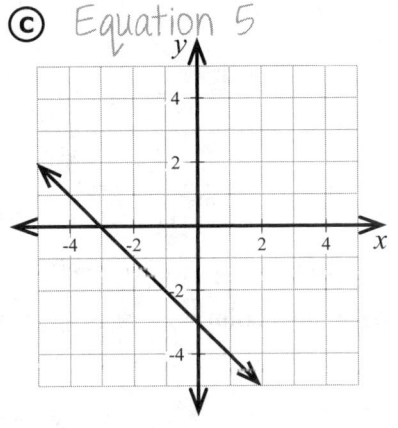

---

⑦ Use the transformation rule $(x, y) \rightarrow (2x, 3y)$ to fill out the table and redraw the shape.

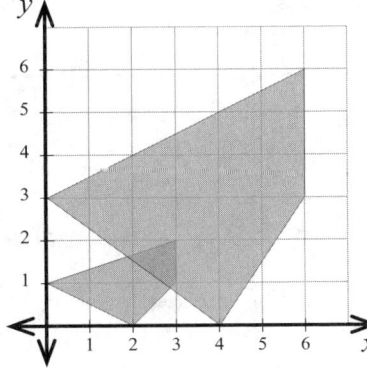

(x, y)	$(2x, 3y)$
$(0, 1)$	$(0, 3)$
$(3, 2)$	$(6, 6)$
$(3, 1)$	$(6, 3)$
$(2, 0)$	$(4, 0)$

8 Use the transformation rule $(x, y) \rightarrow (x + 6, y - 2)$ to move and redraw this graph. Record the points in the table.

(x, y)	$(-2, 4)$	$(1, 3)$	$(0, 1)$	$(-1, 0)$	$(-2, 0)$	$(-3, 2)$
$(x + 6, y - 2)$	$(4, 2)$	$(7, 1)$	$(6, -1)$	$(5, -2)$	$(4, -2)$	$(3, 0)$

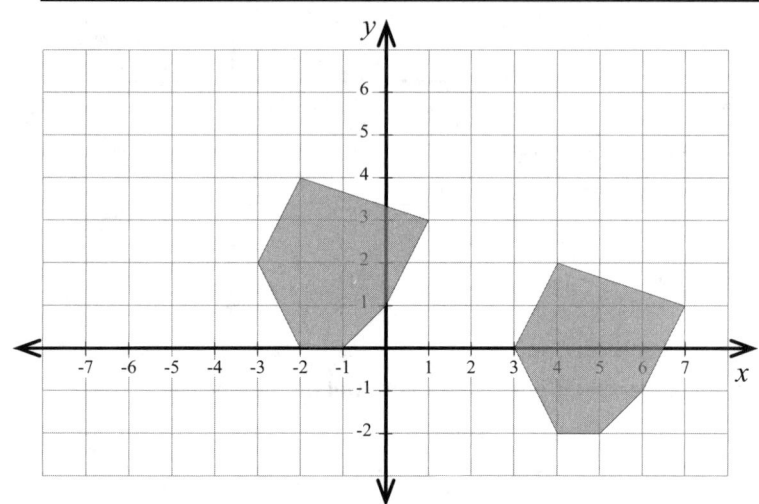

9 Use the $(x, y) \rightarrow (\quad , \quad)$ notation to show the transformation rule for a shape that moves right 4 units and up 1 unit.

$$(x, y) \rightarrow (x + 4, y + 1)$$

10 Kayla's body temperature had been normal all morning, but it began rising steadily after noon. Which graph shows her temperature during the day?

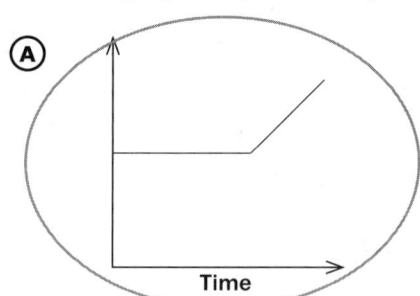

Time

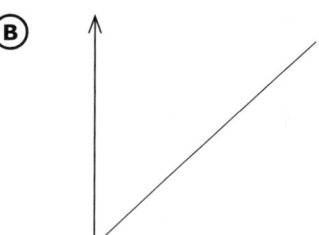

Time

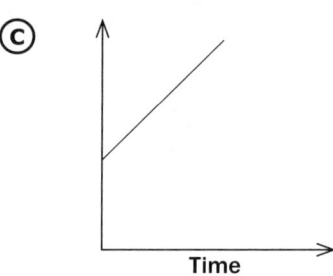

Time

11 Find solution points for $y = 1 - x$.

a Find a solution point if $x = 0$.

$(0, 1)$

b Find a solution point if $x = 2$.

$(2, -1)$

c Find a solution point if $x = -3$.

$(-3, 4)$

d Find a solution point if $x = 20$.

$(20, -19)$

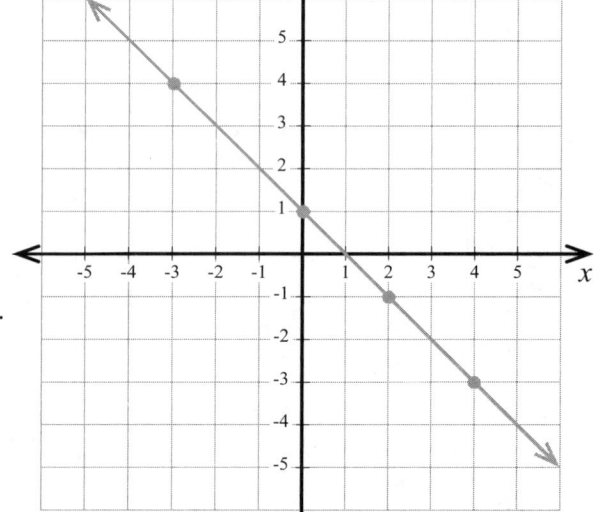

e Circle the solution points and cross out non-solution points.

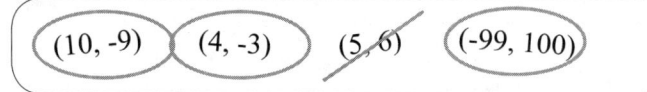

$(10, -9)$ $(4, -3)$ $(5, 6)$ $(-99, 100)$

f Plot the solution points that fit on the graph paper and use them to draw the graph of $y = 1 - x$.

Unit 6

Mental Mathematics

Using approximations to make exact calculations

Students rehearse complements (begun in Units 2 and 3) to increase competence, confidence, and familiarity. They then give special attention to the kinds of adjustments needed in order to use approximations generated by adding and subtracting 10 and 20 as a way of adding and subtracting numbers like 9, 8, 19, and 18. You may find it helpful to write strategy reminders on the board. For instance, in Activity 5, Adding to 8, you might write "$x + 8 = x + 10 - 2$."

In nearly all of these mental mathematics activities, students "enact a function": an input-output rule is established at the outset, and students give the output for each input they hear. Each function rule focuses on a key mathematical idea or property (e.g. complements or the distributive property) that students begin to feel intuitively.

After introducing the day's task, the teacher deliberately does not reiterate the task but says only the input numbers for students to transform. Minimizing words lets students focus on the numerical pattern of the activity, helping them perceive the structure behind the mathematics. A lively pace maximizes practice and keeps students engaged.

Distance to 29

If students struggle with Step 1, revisit the scaffolding offered in Unit 2 Activity 4.

Mental Mathematics • Activity 1
Distance to 30, then 29

PURPOSE

Mental mathematics often requires that we use a "round" number approximation and then refine that to get exact answers. The purpose of activities like this one is to help students develop the capacity to keep mental track of such multi-step processes. Students continue to build working memory and hold many ideas in mind at once.

Introduce:
"I'll say a number, and you tell me the distance from that number to 30. So if I say 12, you say … yes, 18. This may feel too easy at first, but we'll add challenge later. Ready?"

About this sequence:
Step 1 reviews distance to 30, and in Step 2, students calculate distance to 29, thinking, "Distance to 30, and then adjust appropriately."

Step 1: Begin by asking students to calculate distance to 30. Start with one- or two-digit integers between -30 and 30, and then move to integers and halves between -100 and 100.

25	5	**-10**	40
29	1	**17**	13
20	10	**29½**	½
30	0	**31½**	1½
15	15	**40**	10
14	16	**61**	31
4	26	**-4**	34
1	29	**7**	23
-1	31	**87**	57

Step 2: Have students instead tell you the distance to 29. Use mostly positive integers less than 29. Use fractions and negatives for challenge once students catch on.

25	4	**1**	28
15	14	**-1**	30
29	0	**-10**	39
21	8	**-12**	41
4	25	**28**	1
3	26	**28½**	½
10	19	**30**	1
12	17	**32**	3
15	14	**40**	11
13	16	**41½**	12½

Extension: Continue distance to 29, with inputs between -100 and 100.

0	29	**50**	21
-15	44	**58**	29
-17	46	**84**	55

Mental Mathematics • Activity 2
Distance to 100, then 103

Distance to 103

PURPOSE
Calculating distance to 100 from a given number requires students to keep track of how far the number is from the nearest multiple of 10 and how far that is from 100. Distance to 103 requires a further refinement. This activity continues to develop students' working memory and their capacity to perform a calculation by thinking of it in parts and keeping mental track of the process.

Introduce:
"Today we're finding the distance to 100. I'll say a number, and you tell me how far it is from 100. So if I say 30, you say … yes, 70. 35? Yes, 65. Let's go."

About this sequence:
The adjustment in this activity depends on where the input is relative to 0, 100, and 103. "Difficult" locations (negative numbers and those greater than 103) appear less frequently in the sequence.

> When using distance to 100 as a step toward finding the distance to 103, one cannot just memorize a procedure; the adjustment depends on where the input is relative to 0, 100, and 103. Picturing the location helps.

Step 1: Distance to 100. Use mostly integers, including negatives. Include a half or a decimal if students are ready. For example, if students are comfortable with 75, try $75\frac{1}{2}$.

40	60	-71	171
95	5	33	67
65	35	53	47
15	85	-53	153
10	90	75	25
0	100	$75\frac{1}{2}$	$24\frac{1}{2}$
-10	110	181	81
-20	120	381	281
-29	129	98	2

Step 2: Distance to 103. Use integers. If you include negatives, use mostly multiples of 10.

100	3	0	103
90	13	-5	108
95	8	-50	153
85	18	-60	163
50	53	-10	113
56	47	40	63
75	28	43	60
74	29	43.6	59.4
25	78	105	2
$24\frac{1}{2}$	$78\frac{1}{2}$	150	47
41	62	153	50

Extension: Distance to 103 with integers greater than 200.

203	100	325	222
253	150	404	301
250	147	401	298

Distance to 403

Mental Mathematics • Activity 3
Distance to 400, then 403

PURPOSE

Students build on their ability to coordinate groups of tens and individual units, with the extra challenge of keeping track of hundreds. To do this, students must mentally juggle more information and skills than they have previously been asked to and further stretch their working memory.

Introduce:

"We are calculating the distance to 400 today. So if I say 100, you say … yes, 300. How about 150? Yes, 250. Let's go."

About this sequence:

With positive and negative inputs, students must change strategies (e.g. *subtracting from* 400 vs. *adding to* 400, though students may not think of either as subtraction). This has come up in previous "Distance to" activities. Scaffold, at first, by including 0 as a transition.

Picturing roughly where a number "lives" on the number line helps with these calculations. Is the number to the left of 0? Is it to the right of 400? Is it between 0 and 400? Is there a convenient "pit stop" between the input and 400?

Step 1: Have students find distance to 400. Use 2- and 3-digit integers between -1000 and 1000.

200	200	**185**	215
150	250	**187**	213
10	390	**300**	100
90	310	**260**	140
75	325	**262**	138
25	375	**500**	100
0	400	**-500**	900
-25	425	**-525**	925
-75	475	**330**	70
180	220	**338**	62

Step 2: Use similar inputs to Step 1, but have students calculate the distance to 403.

100	303	**-220**	623
300	103	**10**	393
301	102	**15**	388
320	83	**115**	288
323	80	**315**	88
200	203	**275**	128
0	403	**175**	228
-10	413	**150**	253
-100	503	**156**	247
-200	603	**117**	286

Extension: Allow inputs greater than 403.

410	7	**475**	72
450	47	**600**	197
455	52	**605**	202

Mental Mathematics • Activity 4
Adding 9

PURPOSE

The next few activities involve adding and subtracting numbers that are close to multiples of 10, extending and strengthening students' ability to use round numbers and then adjust. In this activity, students warm up by adding 10, and then switch to adding 9. This mental shortcut for adding "difficult" numbers will be extended in later activities in which students combine this shortcut with others.

Introduce:

"We are doing addition today. So if I asked you to add 10 to 43, you would say 53, right? But if I asked you to add 9 to 43, what would you say? Yes, one less: 52. Let's try some more. "

Step 1: Add 10.

23	33	**82**	92
83	93	**67**	77
49	59	**467**	477
59	69	**66**	76

Step 2: Add 9. Say: "Adding 9 is adding 1 less than adding 10. From now on, add 9 to whatever number I give you."

25	34	**84**	93
85	94	**68**	77
41	50	**468**	477
51	60	**60**	69

Step 3: Just continue with no new announcement to the class, but use some larger inputs and very simple decimals or fractions in a scaffolded way (e.g. 54, $54\frac{1}{2}$).

50	59	**$71\frac{1}{2}$**	$80\frac{1}{2}$
53	62	**77**	86
75	84	**$77\frac{3}{4}$**	$86\frac{3}{4}$
78	87	**68.3**	77.3
38	47	**500**	509
11	20	**512**	521
65	74	**961**	970
$65\frac{2}{3}$	$74\frac{2}{3}$	**965**	974
90	99	**$165\frac{1}{2}$**	$174\frac{1}{2}$
92	101	**188**	197

Extension: Include negatives.

0	9	**-18**	9
-9	0	**-50**	-41

- It often helps to say the interim step (adding 10) softly to oneself before calling out a response.
- Remember that this and subsequent activities require at least two steps. Expect slight pauses while students think, and even whisper to themselves, those intermediate steps.

Mental Mathematics • Activity 5
Adding 8

PURPOSE

The purpose of this activity is to extend and help students generalize the idea of using an "easy" number to approximate and then using an adjustment to get an exact answer. Students get a gut feeling for the algebraic equation $x + 8 = x + 10 - 2$.

Introduce:

"If I say 45 + 10, you would of course say 55. If instead I say 45 + 8, what would you say? Yes, exactly 2 less: 53. Let's try some more."

Step 1: Add 10.

74	84
45	55
37	47
134	144

Say, "Got it? Now, I want you to add 8 to anything I say. Ready?"

75	83
42	50
39	47
24	32
124	132

Step 2: Just continue with no new announcement to the class, but use some larger inputs and very simple decimals or fractions in a scaffolded way (e.g. 24, 24.3).

23	31	**8**	16
32	40	**28**	36
47	55	**$28\frac{1}{2}$**	$36\frac{1}{2}$
71	79	**68**	76
37	45	**35**	43
2	10	**35.1**	43.1
64	72	**23**	31
26	34	**19**	27
36	44	**0**	8
16	24	**-1**	7

Extension: Include more fractions and/or decimals without scaffolding.

9	17	**$70\frac{2}{3}$**	$78\frac{2}{3}$
9.1	17.1	**$54\frac{1}{4}$**	$62\frac{1}{4}$
27.4	35.4	**61.8**	69.8

Mental Mathematics • Activity 6
Adding 19

PURPOSE
As in the previous two activities, students add a simpler number, then adjust. Learning to remember that it is possible to use approximation followed by refinement takes time. Students will continue developing this habit and skill in the next activity (Adding 18).

Introduce:

"You certainly know how to add 10 to any number I say. You can also easily add 20 to any number I say. Try it: 34 (54), 23 (43), 67 (87). Too easy! But that lets you figure out how to add 19 to anything I say. So that's what we'll do today. What's 75 + 19? Ready?"

About this sequence:

Previous activities have let students develop the connection between adding a multiple of 10 and adding a nearby number. By now, students should be finding their own methods for approximating and then adjusting.

- Let students know that it is okay to whisper intermediate answers to themselves. For example, to add 47 + 19, they might think 47 + 20 and *say* 67 (to themselves) before adjusting the answer to 66 and saying that out loud. Most people who excel at mental mathematics do exactly that!
- Expect the short pauses as students think or whisper to themselves.

Step 1: Begin with positive integers less than 100.

47	66	**26**	45
27	46	**55**	74
32	51	**71**	90
49	68	**93**	112
29	48	**82**	101
26	45	**56**	75
34	53	**29**	48
38	57	**37**	56
15	34	**46**	65
21	40	**63**	82

Step 2: Allow inputs to exceed 100.

144	163	**424**	443
113	132	**454**	473
181	200	**77**	96
129	148	**777**	796
200	219	**83**	102
212	231	**42**	61
56	75	**64**	83
48	67	**184**	203
548	567	**492**	511
67	86	**499**	518

Extension: Include fractions and/or decimals.

0.5	19.5	$25\frac{1}{2}$	$44\frac{1}{2}$
13.1	32.1	$44\frac{2}{3}$	$63\frac{2}{3}$
19.9	38.9	$32\frac{3}{4}$	$51\frac{3}{4}$

Mental Mathematics • Activity 7
Adding 18

PURPOSE
Students continue to practice mental addition with "difficult" numbers (e.g. 18) by approximating with a simpler number (e.g. 20) and then adjusting.

Introduce:

"Yesterday, you figured out how to add 19 to anything at all. I bet you'll do just as well adding 18 to anything! Ready?"

- If necessary, remind students that almost everybody, no matter how "good" they are at mental mathematics, finds it easier to keep track of computations like this by whispering intermediate answers to themselves.
- Expect (and allow) the short pauses as students think, but, to keep the pace lively, be ready with the next input as soon as they respond.
- If you like, you might write $x + 18 = x + 20 - 2$ on the board and ask your students how the activity demonstrates the equation.

Step 1: Begin with positive integers less than 100.

8	26	25	43
4	22	54	72
18	36	7	25
27	45	94	112
34	52	8	26
46	64	16	34
64	82	56	74
58	76	87	105
14	32	44	62
29	47	65	83

Step 2: Include inputs that exceed 100.

14	32	315	333
25	43	68	86
12	30	104	122
15	33	39	57
115	133	75	93
125	143	34	52
4	22	39	57
17	35	73	91
117	135	474	492
13	31	453	471

Extension: Include fractions and decimals.

3.7	21.7	$75\frac{1}{3}$	$93\frac{1}{3}$
18.1	36.1	96.4	114.4
$18\frac{3}{4}$	$36\frac{3}{4}$	$47\frac{1}{2}$	$65\frac{1}{2}$

Mental Mathematics • Activity 8
Subtracting 9

PURPOSE

Students have worked extensively with addition, but in this activity, they reverse the process and subtract (in this case, they subtract 9 by subtracting 10, and then adding 1). Most people find this much more difficult; it takes more concentration because the direction in which one must adjust the approximated answer is not as intuitively felt.

Introduce:

"Subtracting 10 from anything is too easy for you, but warm up with it for a moment and then we'll make it more challenging. Any number I say, you subtract 10: 35 (25), 85 (75), 64 (54), 47 (37), 147 (137), 747 (737). Yup. Too easy! So this time, subtract only 9 from any number I say!"

Step 1: Use positive integers less than 100 to warm up.

36	27	**25**	16
31	22	**15**	6
43	34	**86**	77
79	70	**11**	2
9	0	**46**	37
85	76	**28**	19
73	64	**58**	49
20	11	**70**	61
34	25	**62**	53
35	26	**32**	23

Step 2: Introduce integers greater than 100.

17	8	**211**	202
117	108	**65**	56
417	408	**167**	158
427	418	**367**	358
454	445	**114**	105
81	72	**84**	75
33	24	**95**	86
206	197	**327**	318
143	134	**219**	210
59	50	**983**	974

- Expect short pauses before answers from students, but be ready with a new input when they do respond.
- If students have difficulty subtracting 9, you may wish to create a step in which students first subtract 10, and then subtract 9 to from the same number (e.g., 78 – 10? 78 – 9?).

Extension: Include decimals and fractions.

87.1	78.1	**$112\frac{1}{4}$**	$103\frac{1}{4}$
$64\frac{1}{2}$	$55\frac{1}{2}$	**45.6**	36.6
$71\frac{2}{3}$	$62\frac{2}{3}$	**99.4**	90.4

Mental Mathematics • Activity 9
Subtracting 8

PURPOSE

Students continue developing the inclination and skill to approximate with "nice" numbers and then adjust the answer. One way to approximate subtracting 8 is to subtract 10, but because that subtracts too much, students must adjust by *adding* to the result. Students may quickly understand intuitively what they need to do, but it takes concentration to keep track and it's easy (and very normal!) to trip up. With time, we get better, but this is never easy.

Introduce:

"We know that subtracting 10 from any number is trivial for you. You've used that to figure out how to subtract 9 from any number. Today, let's try subtracting 8. This takes concentration. Remember, you can whisper parts of the answer to yourself. Ready?"

Step 1: Begin by using 2-digit integers between 0 and 100.

44	36	**54**	46
17	9	**34**	26
31	23	**98**	90
78	70	**76**	68
45	37	**59**	51
11	3	**36**	28
19	11	**34**	26
40	32	**89**	81
66	58	**43**	35
32	24	**51**	43

Step 2: Introduce 3-digit numbers, decimals, and fractions.

78	70	**89**	81
90	82	**90**	82
121	113	**90.8**	82.8
76	68	**86**	78
65	57	**322**	314
233	225	**99**	91
298	290	**187**	179
30	22	**62**	54
30½	22½	**130**	122
43	35	**128.5**	120.5

Extension: Subtract 10 from negative numbers.

0	-10	**-49**	-59
-6	-16	**-21**	-31
-13	-23	**-99**	-109

The extension is not difficult arithmetic but is a big change in thinking. For intrepid students, you might pose it as a brief new challenge.

Mental Mathematics • Activity 10
Subtracting 19

PURPOSE

In this activity, students subtract 19 by using any mental shortcut they like, such as subtracting 20 and then adjusting, or subtracting 9 and then subtracting 10. The purpose is to support students in being flexible with their methods and developing the capacity to manage those methods mentally.

Introduce:

"Subtracting 20 from any number is probably too easy for you now, but let's warm up with it anyway. For example, subtract 20 from 35 (15), 47 (27), 91 (71), 76 (56). Perfect! So now let's try subtracting 19. This takes concentration. Remember, it's okay to whisper to yourselves before you call out an answer. Ready?"

- The arithmetic is not hard, but the level of concentration needed in this task can be exhausting! Watch your students and gauge when it is time to stop. You want them to feel really competent at the end, and not yet worn out.
- If students are doing well but looking stressed, end Step 2 early. If they are not doing well, go back to subtracting 20 for a bit. Even if students do well with all the hard problems, don't prolong this practice.

Step 1: Use integers between 19 and 100 to start.

48	29	**54**	35
70	51	**53**	34
35	16	**52**	33
26	7	**82**	63
49	30	**38**	19
86	67	**45**	26
93	74	**91**	72
63	44	**87**	68
32	13	**53**	34
19	0	**23**	4

Step 2: Introduce 3-digit numbers, and include the occasional decimal or fraction.

48	29	**67**	48
92	73	**263**	244
24	5	**91**	72
78	59	**32**	13
52	33	**32.1**	13.1
152	133	**33.1**	14.1
135	116	**33.5**	14.5
46	27	**533**	514
295	276	$36\frac{1}{2}$	$17\frac{1}{2}$
21	2	$136\frac{1}{2}$	$117\frac{1}{2}$

Extension: Subtract 20 from negative integers.

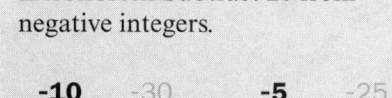

-10	-30	**-5**	-25
-18	-38	**-43**	-63
-50	-70	**-58**	-78

Mental Mathematics • Activity 11

Subtracting 18

PURPOSE

Students practice breaking up seemingly complicated calculations into simpler steps (in this case, subtracting 18 by first subtracting 20 and then adjusting). Students continue to challenge their working memory and skill at tracking their progress.

Introduce:

"Since we've figured out how to subtract 19 from anything, let's do the same with 18. First let's warm up by subtracting 20. Subtract 20 from 56 (36), 72 (52), 91 (71), 36 (16). Yes, of course. Now let's subtract 18."

- These problems take a lot of concentration. You might model how to keep track of such computations by whispering partial answers to yourself as you do one.
- To keep the pace lively, be ready with the next input as soon as students respond, but don't pressure them to be fast; **the goal is not speed but competence at this way of thinking.** Expect (and allow) the short pauses as students think.

Step 1: Use integers between 18 and 100.

83	65	**54**	36
25	7	**84**	66
64	46	**38**	20
73	55	**56**	38
36	18	**39**	21
91	73	**96**	78
41	23	**43**	25
51	33	**89**	71
54	36	**47**	29
81	63	**61**	43

Step 2: Include 2- and 3-digit numbers, including some decimals and fractions.

36	18	**181**	163
66	48	**180.9**	162.9
120	102	**230.7**	212.7
83	65	$30\frac{1}{2}$	$12\frac{1}{2}$
121	103	**24**	6
87	69	**267**	249
325	307	**360**	342
367	349	$362\frac{1}{4}$	$344\frac{1}{4}$
492	474	$352\frac{3}{4}$	$334\frac{3}{4}$
43	25	**97**	79

Extension: Include negative integers.

0	-18	**-61**	-79
-10	-28	**-43**	-61
-25	-43	**-88**	-106

Mental Mathematics • Activity 12
Subtracting 99

PURPOSE

The purpose of this activity is to let students notice that their methods are general and can be applied in situations that they haven't encountered before. At this point, students should have a principle pretty well established: to subtract 99, first do something simpler, and then adjust.

Introduce:

"Subtracting 100 from any big number is already too easy for you, but let's warm up with that. Subtract 100 from 450 (350), 321 (221), 775 (675). Of course! So let's try subtracting 99 instead."

Step 1: Use integers greater than 100.

148	49	**424**	325
213	114	**580**	481
427	328	**533**	434
354	255	**436**	337
142	43	**227**	128
285	186	**331**	232
391	292	**797**	698
279	180	**920**	821
161	62	**843**	744
163	64	**642**	543

Step 2: Introduce simple fractions and decimal numbers.

116	17	**274**	175
113	14	**800**	701
272	173	**$383\frac{3}{4}$**	$284\frac{3}{4}$
272.8	173.8	**346**	247
462	363	**527**	428
805	706	**$731\frac{2}{3}$**	$632\frac{2}{3}$
491	392	**978**	879
719	620	**290.4**	191.4
611	512	**143**	44
$611\frac{1}{3}$	$512\frac{1}{3}$	**143.1**	44.1

You might help students generalize by asking them how this activity relates to their work subtracting 19.

Extension: Subtract 100 from integers less than 100.

70	-30	**87**	-13
67	-33	**14**	-86
44	-56	**33**	-67

Subtract 100

Mental Mathematics • Activity 13
Subtracting or adding 100, crossing 0

You may wish to ask students how this activity reminds them of the Distance to 100 activity in Unit 2.

PURPOSE

Students have practiced $100 - n$ where $0 \leq n \leq 100$ in Unit 2 Activity 11. Now, they are computing $n - 100$ for those same numbers, and $n + 100$ for the corresponding negative numbers. The purpose of this activity is to let students sort out how to make these adjustments and how to manage calculations that cross zero.

Introduce:

"We will start by subtracting 100 today. So if I say 100, you say … of course, 0. What about 90? (-10) Yes, and 10? (-90) Good, and 70? (-30) Yes! 20? (-80) 50? (-50) Perfect! Keep going."

Step 1: Have students subtract 100 from integers less than 100.

40	-60	**50**	-50
60	-40	**55**	-45
59	-41	**20**	-80
58	-42	**26**	-74
20	-80	**20**	-80
21	-79	**18**	-82
90	-10	**70**	-30
92	-8	**71**	-29
90	-10	**72**	-28
88	-12	**99**	-1

Step 2: "This time *add* 100 to whatever number I give you."

-60	40	**-50**	50
-40	60	**-45**	55
-41	59	**-80**	20
-42	58	**-74**	26
-80	20	**-20**	80
-79	21	**-22**	78
-10	90	**-70**	30
-8	92	**-69**	31
-10	90	**-68**	32
-12	88	**-1**	99

Extension: Subtract 98.

32	-66	**71**	-27
11	-87	**89**	-9
46	-52	**65**	-33